Common Insect and Mite Galls of the Pacific Northwest

STUDIES IN ENTOMOLOGY NUMBER FIVE

By
Hiram Larew
Joseph Capizzi

Oregon State University Press
and
Oregon State University
Extension Service,
College of Agricultural Sciences
Corvallis, Oregon

The paper in this book meets the guidelines for permanence and durability of the Committee on Production Guidelines for Book Longevity of the Council on Library Resources.

Library of Congress Cataloging in Publication Data

Larew, Hiram, 1953-
Common insect and mite galls of the Pacific Northwest.

(Oregon State monographs. Studies in entomology; no.5)
1. Galls (Botany)—Northwest, Pacific—Identification.
2. Galls (Botany)—Identification. I. Capizzi, Joseph,
1924— . II. Title. III. Series.
QL 461.07 no. 5 [SB767] 595.7'09795s 83-2321
ISBN 0-87071-055-9 [634.9'62'09795]

ACKNOWLEDGEMENTS

Helpful suggestions on this publication were made by Frank Smith, Paul Oman, Arthur Antonelli, John Mellott and Allen Fechtig. Photographs of the elm gall, plum leaf curl, maple bladder gall, walnut erineum and Douglas-fir needle gall came from the files of the Oregon State University Extension Service (Entomology). The cover illustration of the bullet gall on Garry Oak is by Wendy Madar.

CONTENTS

INTRODUCTION

Anyone with an interest in plants or insects, anyone who cares for a lawn, grows a crop, or enjoys an outdoor walk sooner or later will notice galls. The purpose of this booklet is to provide brief answers to some of the common questions that people ask about galls and to describe some of the mite and insect galls found in the Pacific Northwest. Undoubtedly, we have left out some galls that you will find, but the ones discussed are those that people notice most frequently. The booklet is also a stepping stone into the technical literature. For those of you interested in finding out more about galls, references that give additional information are cited throughout the text. Any university library should have most of the cited journals. A glossary of technical terms is located at the back of the booklet. We use metric measurements throughout the booklet. Remember: 25.4 millimeters (mm) = 2.54 centimeters (cm) = 1 inch.

QUESTIONS COMMONLY ASKED ABOUT GALLS

What is a gall?

A gall is an abnormal proliferation, growth, or swelling on a plant caused by another organism. With such a broad definition, you should not be surprised to learn that galls include very different kinds of abnormalities. For example, abnormal patches of leaf hairs (erinea) are galls, as are the more familiar ball-like swellings on leaves and stems. The gall-producing organism for a part of its life lives and feeds in the gall. A gall is *not* a plant seed or fruit.

With few exceptions, galls are rarely larger than a fist. They take on various shapes (some oak galls in California look like mushrooms) and are pink, red, white, pale green, yellow, or brown. Do not be surprised if you find galls on flowers, fruits, or roots—all plant parts are susceptible to attack, not just leaves and stems. This is especially true for the oaks—everything from catkins to roots (acorns too!) can be galled.

What organisms cause galls?

Many kinds of organisms cause galls. Viruses, bacteria, slime molds, fungi, nematodes, rotifers, mites, and insects all contain gall-forming members. Although this booklet emphasizes the galls caused by mites and insects, mention should be made of the very important non-arthropod gall formers. For example, the most beneficial gall former is the nitrogen-fixing bacterium, *Rhizobium*, that forms and lives in nodules (i.e., galls) on the roots of legumes. This gall former greatly improves soil fertility.

On the other hand, another bacterium, *Agrobacterium*, a close relative of *Rhizobium*, causes the serious plant disease, crown gall. Crown gall is responsible for extensive damage to several crops, especially to fruit and nut trees and to berry canes. In the same vein, club root of cabbage, caused by the slime mold, *Plasmodiophora*, galls and thus damages the roots of various crucifers such as cabbage, broccoli, and brussel sprouts. The attacked plants suffer a chronic wilt.

Several gall-forming fungi damage crops. Crown wart (*Urophlyctis*) of alfalfa (photo 1) is caused by a water-mold fungus. Fungal smuts (*Ustilago*) that form galls on the heads of wheat or corn are caused by basidiomycetous fungi. If you grow azaleas, you may know the azalea leaf gall (photo 2) that is caused by another basidiomycete, *Exobasidium*. Peach leaf curl, a common and important gall disease, is caused by the ascomycetous fungus, *Taphrina*. Lastly, root-knot nematodes (*Meloidogyne*) (sometimes called eel worms) are very common gall-forming pests that attack a wide variety of crops.

Photo 1: Crown wart galls (arrow) on alfalfa collected in mid-August. Scale in mm.

Photo 2: Azalea leaf gall (arrow).

In general, most of the gall formers that have an impact on humans attack the roots, crown, or seeds of crop plants. Additionally, most of the agriculturally important gall formers are caused by bacteria, fungi, or nematodes. A few gall-forming mites and insects are agricultural pests, but most are considered unimportant.

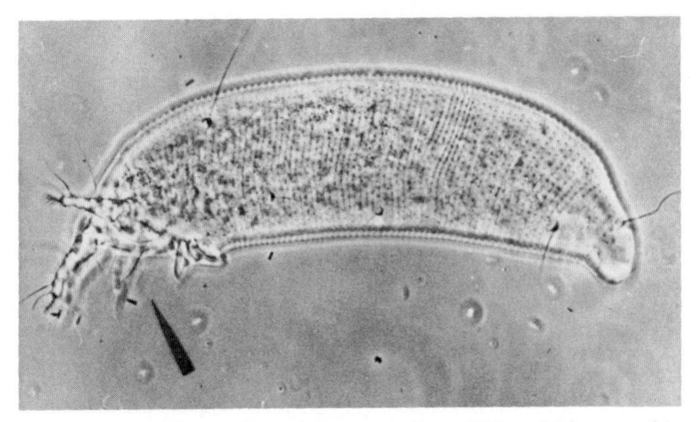

Photo 3: Eriophyoid mite, *Phytoptus avellanae* Nal., which causes big bud galls on filbert trees. The arrow points to the anterior end of the mite where the legs and mouthparts occur.

Gall-Forming Mites. Most gall-forming mites (photo 3) belong to the super order Eriophyoidea, but not all eriophyoids form galls. You really need a hand lens or microscope to see these small mites. Because of their strange appearance, however, they are worth the effort. They have only two pairs of legs (most mites have four pairs), both attached at the animal's front end, and they use the legs to pull or drag their long, worm-shaped bodies. The mites puncture and feed on individual plant cells with their needle-like mouth parts. As a super order, they attack a wide variety of plants, but each species is usually host-specific.

Table 1. Major Groups of Gall-forming Insects Found in the Pacific Northwest

Order	Family	Common Name
Homoptera	Aphididae	Aphids (Plant Lice)
"	Adelgidae	Adelgids
"	Coccidae	Scales
Diptera	Cecidomyiidae	Gall Midges
"	Tephritidae	Fruit Flies
Coleoptera	Cerambycidae	Long-horned Beetles
"	Curculionidae	Weevils
Hymenoptera	Tenthredinidae	Sawflies
"	Cynipidae	Gall Wasps

Gall-Forming Insects. See table 1 for a list of the groups of insects that form galls in the Pacific Northwest. On a worldwide basis, table 1 is incomplete; different gall insects predominate in different areas. For example, in India thrips and gall midges are responsible for many galls. In Australia, scale insects are common gall formers that attack eucalypts (gums). In Europe and the Americas, gall wasps are very common, especially on oaks, and several midge and aphid galls are plentiful as well.

Gall insects either use mandibles (jaws) to chew through gall tissue or stylets (hollow needles) to pierce cells of the gall and to suck up cell sap. As with the gall mites, the insects as a group gall many types of plants, but individual species of insects are usually host-specific.

What do gall insects look like?

There are two basic kinds of insect gall formers. The stylet bearers, such as aphids and adelgids (photo 4), have a small, hollow, needle-like stylet attached to their heads. It usually lies along the underside of the insect when not in use. Aphid and adelgid galls often contain several life stages, such as eggs, nymphs, and winged and wingless adults.

Other common gall formers such as gall midges, sawflies, and gall wasps use mandibles to feed on the gall tissues. Usually the only life stage found in these galls is the larva. Sawfly larvae look like caterpillars, and the adults look like broad-waisted wasps. Gall wasp larvae are usually small white legless grubs with brown mandibles; adults (photo 5) are small stingless wasps. Mature gall midge larvae are usually easy to identify. They are orange or white and have a brown T- or Y-shaped sternal spatula

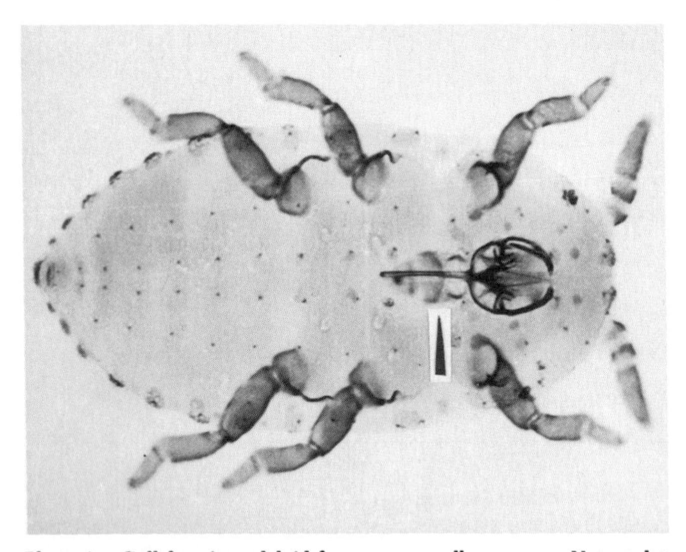

Photo 4: Gall-forming adelgid from anasas gall on spruce. Note stylet mouth parts (arrow).

(photo 6) on their undersurface near their head. Adult midges resemble small mosquitoes.

How are insect and mite galls formed?

The gall insect or mite injects salivary substances into young plant tissues that affect the development of the tissues. Individual plant cells respond to the injection by enlarging and/or

Photo 5: Adult cynipid wasp (*Andricus quercus-californicus* var. *spongiolus* Gill.). Scale in mm.

Photo 6: Cecidomyiid larvae, one clearly showing its sternal spatula (arrow).

dividing—two responses that give rise to the galls. Unfortunately, the chemical composition of the injection and how it controls plant cell responses in unknown.

We do know, however, that the hormone system of the attacked plant organ is affected during gall formation. We also know that the continual presence of the gall former is required if the gall is to reach maturity, because gall growth stops if the gall former is killed or removed. Some gall formers move about during the early stages of gall development. We suspect that this wandering and feeding determines the shape of the gall.

What are the advantages to gall living?

Galls provide a close-at-hand and, in some cases, enriched diet for the gall former. Galls also probably buffer environmental fluctuations and provide a humid cavity in which to develop. They protect the gall former from wind, rain, radiation, and (perhaps) general predators and parasites.

Many galls, however, do *not* protect gall insects from specialized enemies such as parasitic wasps. In fact, we often speak of the gall "community"—a term that refers to the group of parasites, hyperparasites (parasites of parasites) and inquilines ("guests") that invade the gall and often directly or indirectly kill the gall former. Because of high rates of parasitization, there is the good chance that the insect(s) you find in a gall or that emerges from a gall, will not be the gall former.

Do galls harm the plant?

Galls, like fruits, are plant sinks. This means that photosynthates and other plant nutrients are preferentially shunted to the galls and gall formers. Furthermore, galls are probably one-way sinks; the plant can never retrieve much of what goes into a gall. So, by acting as drains, galls undoubtedly harm plants. The fact that gall-forming mites and flies have been used recently as biological control agents to slow the spread of weeds reinforces the argument that arthropod galls are *not* benign.

Many fungal and bacterial galls do considerable damage to crops. Interestingly, however, few gall-forming mites and insects are considered pests. This is because these arthropods usually attack non-crop plants, occur in fairly low numbers, cause little vascular blockage, and/or sap relatively small amounts of nutrients from the large host plant.

One example, however, of an insect gall former that in times past was of tremendous importance is the grape phylloxera, *Phylloxera vitifoliae* (Fitch), an aphid-like insect. *Phylloxera* insects gall both the leaves and roots of grapes. After being accidentally introduced from the U.S. in the mid-1800s, they destroyed many French vineyards. Not until resistant rootstocks from the U. S. were imported was the European wine industry able to recover.

Are galls of any use to humans?

Historically, certain oak galls with high contents of tannins were used to produce permanent inks for legal documents. Since medieval times, oak gall tannins have been used to tan hides and to treat illnesses. Presently, gall formers are useful as biological control agents of weeds. Also, those of you who enjoy figs can thank a small gall-forming wasp (*Blastophaga psenes*) for pollinating and galling the unusual fig flowers. The quality of edible figs would be much lower without these gall insects.

To envision the future importance of galls, we turn to laboratories, because when we understand how gall formers manipulate plant cells, we will have a powerful tool with which to improve plant characters. Also, it is conceivable that someday we will develop galls that can be used as food supplements for livestock—perhaps even for humans.

What do you see when you slice through a gall?

What you see depends on the type of gall. Many galls, for example, contain a single central chamber (unilocular galls) in which one to many larvae, nymphs, and/or adults occur. Other galls contain many chambers (plurilocular galls), each with one to several insects. Galls range in structural complexity from the fairly simple, disorganized, callus-like growth caused by crown gall bacteria to wasp galls with concentric layers of well-organized tissues. Depending on the type of gall and its age, you may have trouble cutting through it. As they mature, many galls (particularly midge and wasp galls) develop a band of thick-walled cells that reinforces the gall wall.

With the exception of some aphid, adelgid, and scale galls, the chamber in which a gall insect or mite lives is lined with succulent cells. These food cells (photo 7), like those cells at the growing tips of plants, have a dense cytoplasm and are rich in proteins, carbohydrates and, in some cases, oils. So, many gall formers feed for part of their lives on enriched plant cells.

How long have galls existed?

Although older examples may exist, we know of a 10- to 25-million-year-old gall of a *Sequoia* cone from Germany that still contains gall midge larvae. There is nothing new about the gall-forming habit. To give you some idea of what a fossil gall looks like, we have included a photograph (photo 8) of a 10- to

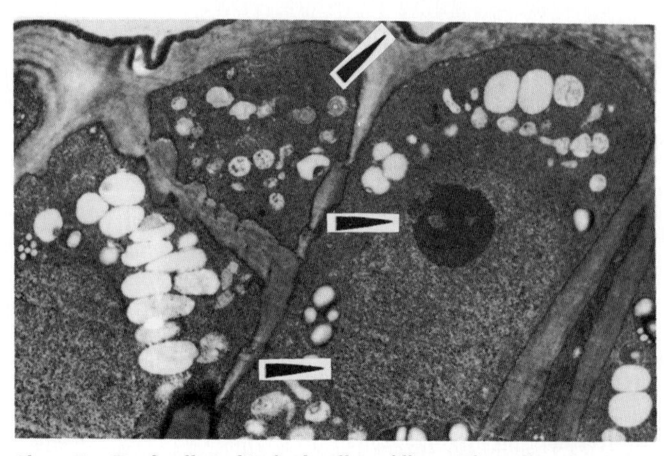

Photo 7: Food cells in big bud gall on filbert. The gall cavity is just above upper arrow. These cells have a dense cytoplasm, a large nucleus (lower arrow) and a prominent nucleolus (middle arrow).

15-thousand-year-old acorn that was galled by wasps; it was taken from the skull of a sabre-toothed cat found in the La Brea Tar Pits.

What do I do with a gall once I have collected it?

There is always the chance that the gall you collect has not been described (i.e., reported in a scientific journal). For that reason, we encourage you to send specimens that you have questions about to your local Extension agent. The more information about the gall you can provide, the better. You should, for example, include when and where you found it. If you know the name of the plant on which you found the gall, include it. Even if you know the plant, the Extension agent will appreciate receiving some ungalled leaves, stems, and (especially) flowers or fruits.

When delivering the gall to your agent, it is best to put a tag with this information and the gall together in a jar or bag. To delay wilt and drying, you can wrap the gall in a moist paper towel and refrigerate it in a sealed jar or plastic bag.

How do I control gall insects?

Recommendations vary. The best policy is to ask your Extension agent. If the infestation is small, you may be able to control it by cutting out the galls and burning them. If the damage is unsightly or heavy, chemical control may be necessary. Timing of chemical application is important; the life stages of the insect that live outside the gall are those most effectively controlled by topical sprays. Your Extension agent can help you decide when it is best to spray.

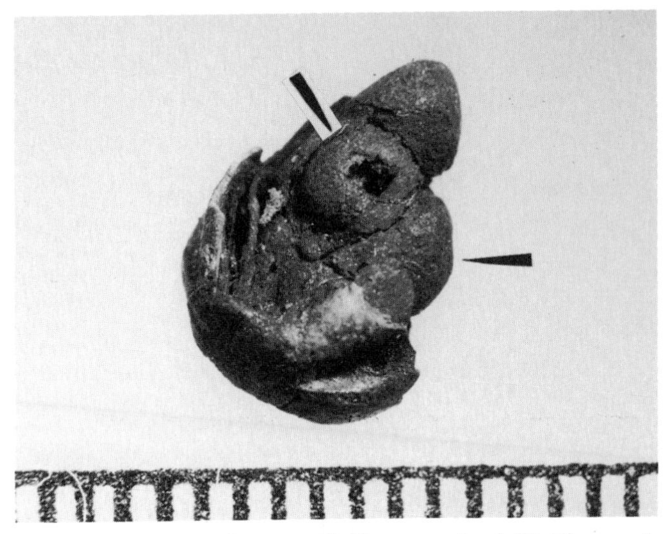

Photo 8: Fossil acorn that was galled by a wasp (cynipid). The acorn's shell has been removed so that the nut meat bearing spherical blisters (arrows) is seen. A single larva developed in each hollow blister. Scale in mm. (Specimen courtesy of the G. C. Page Museum, Los Angeles, California.)

For general information about galls see:

Felt, E. P. *Plant Galls and Gall Makers*. New York: Hafner Press, 1965. Although clumsy, this is still the first place to turn when you need to identify a gall.

Furniss, R. L. and V. M. Carolin. *Western Forest Insects*. U. S. Forest Service Miscellaneous Publication no. 1339 (1980). Many of the gall insects that attack forest trees and shrubs in the West are described.

Johnson, W. T. and H. H. Lyon. *Insects That Feed on Trees and Shrubs*. Ithaca, N. Y.: Comstock Publishing, 1976. This includes excellent photographs of many of the galls on ornamental shrubs.

Keifer, H. H. and others. *An Illustrated Guide to Plant Abnormalities Caused by Eriophyid Mites in North America*. USDA Handbook 573 (1982). Excellent photographs will help you identify mite-caused galls.

Mani, M. S. *Ecology of Plant Galls*. Monogr. Biolog. vol. 12. The Hague: Dr. W. Junk Publ., 1964. This will not help you identify galls, but it is the best general discussion of galls available in English.

Russo, R. A. *Plant Galls of the California Region*. Pacific Grove, Calif.: Boxwood Press, 1979. This has very good, up-to-date coverage of California galls, several of which occur in the Pacific Northwest.

Schuh, J. and D. C. Mote. *Insect Pests of Nursery and Ornamental Trees and Shrubs in Oregon*. Oregon Agricultural Experiment Station Bulletin 449 (1948). An old but useful reference.

Speckled Gall

Host: *Quercus garryana* Dougl.

Gall Former: Gall Wasp (Cynipid). *Besbicus (Cynips) mirabilis* Kinsey (agamic generation)

Any trail through an oak grove is sure to be littered with these light brown, speckled galls (photo 9). Some people call these "pop balls" because they pop under foot. Others call them oak apples or oak cherries. These galls are so common in western Oregon and Washington that we, like the British, should celebrate May 29 as Oak Apple Day.

The galls, often several to a leaf, are attached on the lower leaf surface to the midvein or to one of the major secondary veins. They often remain attached to the leaf even after the leaf has died and fallen. This gall occurs only on Garry oak and has been found over the plant's entire range (British Columbia to northern California). The galls, however, do not occur on every Garry oak; often one tree will be heavily attacked while its neighbor is free of galls. This may indicate that even though all Garry oaks belong to a single species, individual trees differ in their susceptibility to the wasp. This may be because of differences in plant chemistry, phenology or location.

Photo 9: Speckled galls on Garry oak leaves collected in mid-September. Scale in mm.

The young gall (photo 10) originates just under the surface of the midvein. As the gall grows, it ruptures the midvein's surface, and emerges as a tuft of white (sometimes pinkish) hairs. The tufts first appear in early-to-mid-June. A very young gall is solid, and it remains so until it reaches 2 to 3 mm in diameter in late June. From then on, the gall becomes more and more spongy and finally almost hollow. While it grows, the gall is yellowish green and has a velvety covering of short hairs. At maturity the gall is about 2.5 cm in diameter and is shiny reddish brown with light and dark brown speckles. Throughout its development, the gall is spherical except at the pinched point of attachment.

When you cut a mature gall open (photo 11), you see that the outer speckled wall is very thin. A small spherical capsule is suspended in the center of the gall by long white or brown fibers (in young or mature galls, respectively) that radiate from the capsule to the outer gall wall. The growth and separation of these fibers cause the transition from solid to hollow as the gall matures. This gall has one larva and a single chamber. The central capsule in which the larva lives is fleshy while the larva matures, but it becomes brown and brittle in autumn. The food cells that line the inner surface of the capsule are enriched. The larva, like all other gall-wasp larvae, has a blind gut (no anal opening) and excretes no fecal matter until pupation.

According to one account (Evans, 1967), only about 4 percent of the larvae survive to adulthood because most are attacked by parasitic wasps. You may also often find a moth caterpillar,

Photo 10: Young speckled galls (arrows) on Garry oak collected in early June.

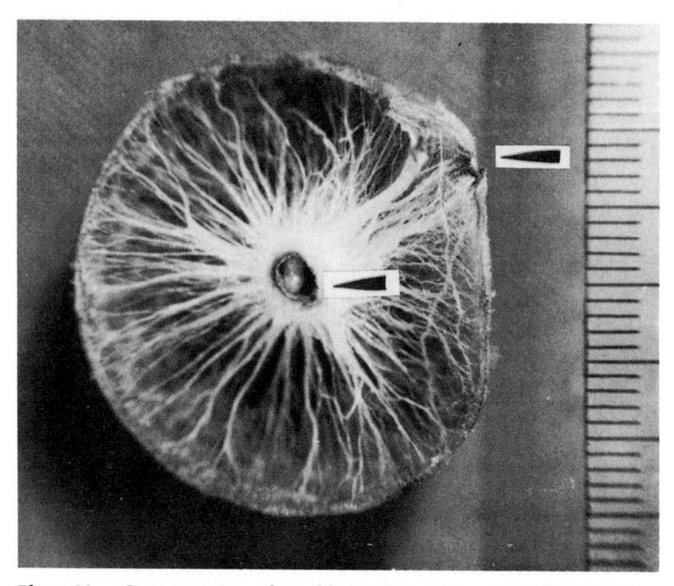

Photo 11: Cutaway view of speckled gall on Garry oak collected in late September. The gall has been removed from the leaf. It was attached to the leaf at the upper arrow. The wasp larva is in the capsule (lower arrow) that is suspended in the center of the gall. Scale in mm.

Olethreutidae: *Melissopus latiferreanus* (Wals.), and its fecal pellets inside the gall. The caterpillar eats just about everything in the gall and, in so doing, kills the gall former.

This gall is but one of the two caused by *B. mirabilis* on Garry oak. Like many other gall wasps, this species exhibits an alternation of generations. The speckled gall is caused by larvae that develop into agamic (asexual) females. Thus, the larvae belong to the agamic generation, and the gall is a product of the agamic generation. Agamic females (all agamics are females) emerge from the speckled gall in December to February and lay eggs (no mating required) in Garry oak leaf buds. The larvae that hatch from these eggs form inconspicuous galls in the buds. The adults that emerge (April to June) from the bud galls are either male or female; thus, they constitute the bisexual generation. After mating, a sexual female completes the cycle by laying its eggs in the midveins, and a speckled gall is produced. We can only guess as to the advantages of such a complex life cycle.

Interestingly, many other species of gall wasps on other American and European oaks form galls that are designed much like the speckled gall. We have no idea why the "suspended capsule" design is such a popular one.

A final point of interest: A. C. Kinsey, the human sexologist, did a great deal of work with this and other wasp galls when he was young. The book that resulted from his work is a classic and useful reference.

For more information on the Garry oak speckled gall see:

Evans, D. "The Bisexual and Agamic Generations of *Besbicus mirabilis* (Hymenoptera: Cynipidae), and their Associate Insects." *Canadian Entomologist* 99 (1967): 187-196.

Kinsey, A. C. *The Gall Wasp Genus Cynips: A Study in the Origin of Species*. Indiana University Studies vol. 16 (1930), Studies nos. 84, 85, 86.

Bullet Gall

Host: *Quercus garryana* Dougl.

Gall Former: Gall Wasp (Cynipid). *Andricus quercus-californicus* var. *spongiolus* Gill. (Also known as *A. spongiolus* Gill.)

Because of its size, color, shape and solidness, this gall should be called the "green apple" gall (photo 12). This is the largest insect gall in the Pacific Northwest, often 7.5 to 10 cm long. Although they usually can be seen from a distance, the easiest way to find these galls in summer is to stand under an oak and look up through the leafy branches. The galls are attached to the stem, frequently in groups of two to ten. Some of the galls in a group may be less than an inch in diameter; usually the smaller the gall, the more spherical it is. The large galls are more kidney-shaped. Notice that the gall occurs on two- to four-year-old branches, and that the branch is swollen slightly where the gall is attached. This is a many-chambered gall; all of the chambers are bunched together at the center of the gall or are sometimes clustered near the point of attachment. Each chamber contains a single larva.

The very young yellow-green galls develop within the stem and rupture the stem's surface in early May. They reach mature

Photo 12: Bullet gall on Garry oak collected in early July. Note twig swelling at the point of gall attachment (arrow).

size in midsummer. The pulpy tissue of young galls is juicy and white. As the gall grows, the tissue surrounding each larval chamber hardens; because the chambers abut one another, it becomes very difficult to cut through the cluster of chambers. The pulp of the gall often begins to rot and turn brown in July. The gall's surface is sometimes spotted black from a mold that probably lives on sugary secretions produced by the gall. During the fall and winter the galls turn light brown. Some fall to the ground, but others remain attached to the stems until the following spring. If you break open an old gall, you see a central hardened mass of larval chambers as well as a stringy mass of fibers that run from the stem to the thick-walled chambers. Plant nutrients pass through these fibers to the larval chambers.

The damage caused by this gall has never been assessed, which may suggest that there is no serious damage. The gall is invaded by many parasites and inquilines. We do not know if this gall-wasp species has an alternation of generations.

Bassettia Twig Gall

Host: *Quercus garryana* Dougl.
Gall Former: Gall Wasp (Cynipid). *Bassettia ligni* Kinsey (agamic generation)

This is an unusual gall. In fact, when you see it, you may be surprised that it is even called a gall, because there is no large stem swelling. This is also one of the few wasp galls in the Pacific Northwest that causes considerable twig dieback.

The galls occur on two- or three-year-old twigs, and the girth of an attacked twig is slightly larger than normal (photo 13). In older specimens, you should see small holes along the surface of an attacked twig through which the adult gall wasps emerge. When you peel the bark off a galled stem (photo 14), you see several 2 mm x 1 mm oblong larval chambers, each aligned with its long axis parallel to the stem's. In heavy infestation, however, this alignment breaks down, and the chambers are packed together randomly. Each chamber is sunken slightly into the stem's wood. The chamber wall is a fairly thick membrane of fleshy plant tissue upon which the larva feeds. When found in large numbers, the chambers may girdle the twig. If not the galls, certainly rodents that tear back the bark to feed on the larvae cause girdling. The twig dies from its tip back to the point of girdling.

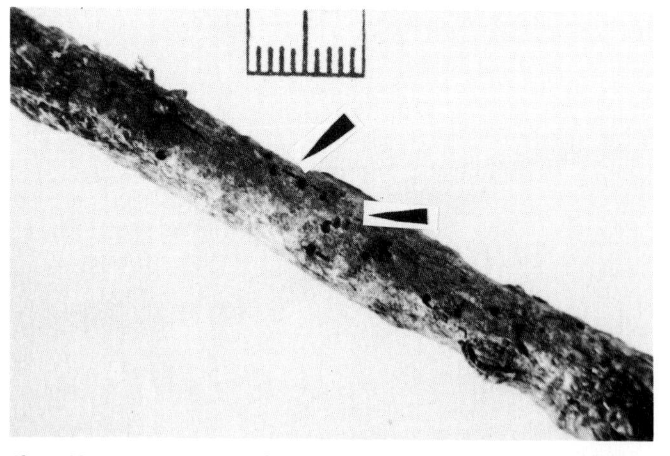

Photo 13: *Bassettia* twig galls on Garry oak collected in late July. Note the puffy appearance of the bark and the numerous adult emergence holes (arrows). Scale in mm.

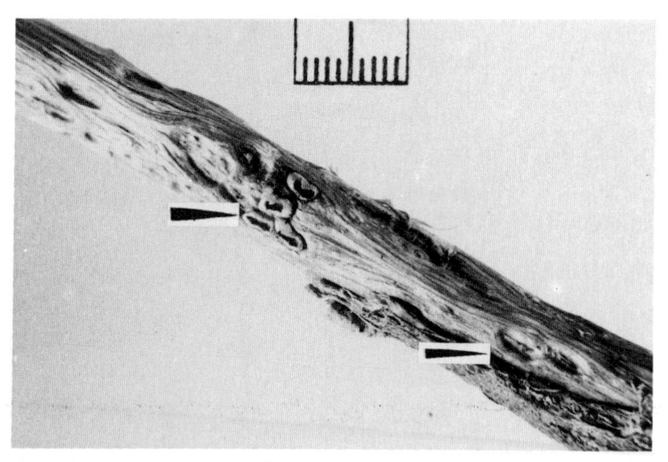

Photo 14: *Bassettia* twig galls on Garry oak collected in late July with bark removed. Cocoon-like larval capsules (upper arrow) lie in depressions in the wood (lower arrow). Scale in mm.

Adult agamic female gall wasps emerge from the twig galls in early spring and lay their eggs in young Garry oak leaves. The resulting galls (the bisexual generation's galls) are fairly inconspicuous leaf blisters from which male and female wasps emerge in June to mid-July. The mated bisexual females lay their eggs in twigs at this time; the resulting twig galls contain pupae by early fall.

For more information on the Garry oak *Bassettia* twig gall see:

Evans, D. "Alternate Generations of Gall Cynipids (Hymenoptera: Cynipidae) on Garry Oak." *Canadian Entomologist* 104 (1972): 1805-1818.

Jumping Gall

Host: *Quercus garryana* Dougl.
Gall Former: Gall Wasp (Cynipid). *Neuroterus saltatorius* Edwards (agamic generation)

This is a very common gall (photo 15) on Garry oak leaves in the Pacific Northwest. The galls are about 1.5 mm long, egg-shaped (Photo 16), and occur on the underside of the leaf. They often occur betwen the major leaf veins. Each gall causes a small yellow spot on the corresponding area of the upper leaf surface. They are easily detached from the leaf. Each gall contains one larva and a single chamber.

A single leaf can bear one to many galls. The galls appear in early spring and remain light green to yellow through midsummer. From July through August, they fall from the leaf. Once on the ground, the galls begin to jump—an action caused by the flexing insect in the gall. If the larva dies early in its development, the gall will turn brown and remain attached to the leaf.

When several galls occur on the leaf, they may cause early leaf death, but we do not know how detrimental this damage is to the plant. Usually the amount of damage caused does not justify large-scale control efforts. The gall wasps are heavily parasitized. In fact, we are not sure if the gall former or a parasite larva does the jumping.

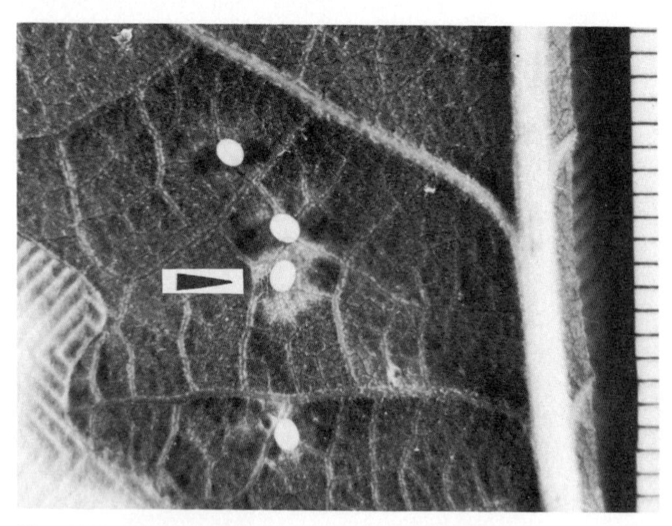

Photo 15: Jumping galls on Garry oak collected in mid-June on the underside of leaves. Note that where the gall is attached, the leaf is yellow (arrow). Scale in mm.

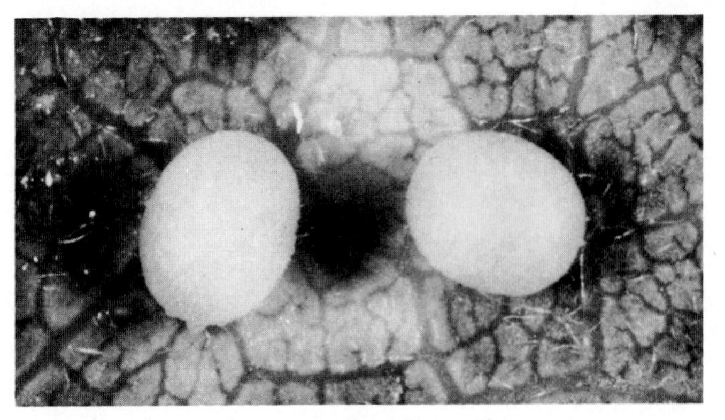

Photo 16: Close-up of jumping galls on Garry oak.

Like many other gall wasps, *N. saltatorius* passes through asexual and sexual generations. Jumping galls contain larvae that will become agamic (asexual) females. The agamic females that emerge from jumping galls in early spring cause small inconspicuous leaf blister galls on Garry oak from which sexual females emerge. Mated sexual females lay eggs in young Garry oak leaves. The resulting larvae form jumping galls.

For more information on the Garry oak jumping gall see:

Ritcher, P. O. and J. Capizzi. *Jumping Oak Galls.* Oregon State University Pest Information Letter no. 12 (1977).

Rosenthal, S. S. and C. S. Koehler. "Heterogony in some Gall-forming Cynipidae (Hymenoptera) with Notes on the Biology of *Neuroterus saltatorius.*" *Annals of the Entomological Society of America* 64, no. 3 (1971): 565-570.

Spherical Stem Gall

Host: *Quercus garryana* Dougl.
Gall Former: Gall Wasp (Cynipid). *Disholocaspis (Callirhytis) washingtonensis* Gill.

This common gall (photo 17) occurs on two- or three-year-old twigs. It ruptures the stem's surface early in spring and is spherical throughout its development. Usually several galls are clustered on a stem. A mature gall is about 8 mm in diameter and may have a short neck where the gall is attached to the stem. A single larva lives in the one large central chamber.

Young galls are dark green and covered with a velvety layer of hairs. Internally, the solid wall is green except for the lining of the chamber, where the food tissue is white. Most galls turn brown and woody in early autumn. The adult wasp emerges in late autumn by chewing a prominent exit hole through the wall. The brown or black empty galls may remain on the tree for one to two years, and frequently groups of young and old galls are found next to each other on a twig.

The wasp and its parasites have not been extensively studied. It is not known, for example, whether the wasp passes through alternate generations.

Photo 17: Spherical stem galls on Garry oak collected in early June.

Other Galls

Host: *Quercus garryana* Dougl.

Several other types of gall wasps attack Garry oaks. In fact, one older source lists 16 different kinds of wasp galls on Garry oak, and recent research has almost doubled that number. The important point is that there is a very good chance that any gall you collect from Garry oak (or from any oak, for that matter) is a wasp gall. With that in mind, two exceptions should be mentioned.

Oak Leaf Erinea

Dense velvety patches of white, yellow-green, or pinkish-to-red hairs (photo 18) occasionally occur on the upper and/or lower surfaces of oak leaves. These erinea, as they are often called, can be collected in late May through late July on fully expanded oak leaves. They are caused by eriophyoid mites that live and feed at the base of the hairs. The hairy patches probably are caused by several different, mostly undescribed species of eriophyoid mites.

Oak Pit Scale (Coccid. *Asterolecanium minus* Lindinger)

The second exception is an important one because of the damage it can cause. At first glance, pit scale damage does not look like a gall, but on examination, one sees that the insect is surrounded by a small (1 to 3 mm in diameter) ring of swollen

Photo 18: An erineum on underside of Garry oak leaf collected in late May.

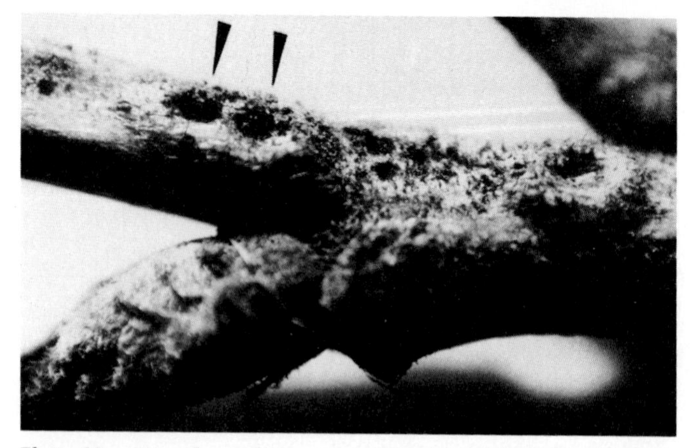

Photo 19: Pit scale on Garry oak collected in early April. The galls (arrows) were formed the previous year. The elliptical pit of swollen tissue contains no insect.

twig tissue (photo 19). This swelling does not cover the insect but forms a pit in which the scale lives and feeds. The insect itself is covered by a thin, brown, hardened plate; this structure is secreted by the insect and is called the scale. Its production by these insects is so distinctive that as a group these are called scale insects.

The female lays eggs under her body in late winter (males of this species are unknown). The newly hatched nymphs crawl away from the female in spring and settle on new or the previous year's twigs. Once settled, the insect remains in place for the rest of its life. The insect feeds with a stylet on the contents of stem cortical cells. Cells of the bark meristem and cortex of the stem divide and enlarge soon after the nymph settles. The characteristic pit develops within a week as these stem tissues swell around the insect. Damage caused by this gall insect may be severe; twig dieback becomes visible in late summer. One commonly finds twigs that are covered by hundreds of scales in abutting pits.

For more information on Garry oak pit scale see:

Koehler, C. S., et al. *Pit Scales on Oak.* University of California Division of Agricultural Science Leaflet no. 2543 (1977).

Parr, T. *Asterolecanium variolosum Ratzeburg, A Gall-forming Coccid, and its Effect upon the Host Trees.* Yale Forestry School Bulletin no. 46 (1940).

Beaked Stem Gall

Host: *Salix* spp.
Gall Former: Gall Midge (Cecidomyiid). *Mayetiola (Phytophaga) rigidae* O.S.

This is a very common stem gall in the Pacific Northwest and throughout much of the U. S. It occurs on several species of willow such as *S. caprea*, *S. muscina*, and *S. cinerea*. It becomes very conspicuous after leaf fall when the swollen stem tips are silhouetted against the sky. Young galls appear in early May just beneath the end of a new twig. At this time, the attacked stem is swollen to twice its normal diameter and the gall is about 2 cm long. A young gall is easily cut open and shows a single tubular chamber that runs the length of the gall. One larva occurs in each chamber.

As the gall grows, it increases five to ten times in diameter. The mature gall in late summer (photo 20) is about 2.5 cm long and 1.5 cm wide at its base. The larva is slow to grow at first, but by late October you can easily find the reddish orange, 3 mm-long larva. Because of the thickness of its woody wall, a mature gall is difficult to cut through. The tip end of the gall tapers to a pointed beak that first turns yellow, then dark brown at maturity. The swollen part of the gall remains green through the fall and

Photo 20: Beaked stem galls (arrows) on willow collected in mid-September. The beaks are brown on the two galls on the right. Scale in mm.

early winter. Young side buds grow from the surface of the gall, and several of these develop as side branches during the course of the summer or the following spring. The gall kills the established stem terminal. Galls often remain on the tree for one to three years, long after the adult midge has left the gall.

Older larvae overwinter in the galls. Pupation occurs in late winter. Adults emerge in early spring through the gall's beak. At least three species of parasitic wasps are known to attack the gall former. Control of the gall former has also been described.

For more information on Willow beaked stem gall see:

Smith, F. F. et al. "Willow beaked-gall midge: Control by insecticides and pruning." *Journal of Economic Entomology* 68 (1975): 392-394.

Wilson, L. F. "Life history and habits of the willow beaked gall midge *Mayetiola rigidae* (Diptera: Cecidomyiidae) in Michigan." *Canadian Entomologist* 100 (1968): 202-206.

Apple Gall

Host: *Salix* spp.
Gall Former: Sawfly (Tenthredinid). *Pontania* spp.

This gall is formed on willows in an unusual way. Instead of larval saliva being responsible for gall formation, a fluid that the female sawfly injects as she lays eggs in the young leaf causes initial gall growth. So, if you open a young gall (photo 21) before the middle of May, you find only an egg in the slit-like gall chamber. It is only later that the larva hatches from the egg and feeds on the awaiting food cells. Most likely, larval saliva is responsible for continued growth of the gall. In October after the leaves have fallen, you can find dead leaves in the litter that bear yellow-green galls (photo 22) containing healthy sawfly larvae. Adult sawflies appear in early spring, and the female uses her saw-like ovipositor to slice young leaf tissue and to deposit her egg in the wound.

The gall contains one larva and is single-chambered. The caterpillar-like larva is white or light purple with a brown head capsule. Galls are spherical to kidney-shaped and usually yellow or yellow-green with several small brown speckles at maturity. Mature galls are 5 to 12 mm long and resemble small green apples attached to the leaf's underside. They protrude slightly through the leaf's upper surface. This upper part of the gall is often red. Galls are usually located next to the midvein, and much of the gall is derived from altered midvein tissue. The wall is solid and wall tissue is white. As the larva feeds, it fills its chamber with reddish-brown fecal pellets.

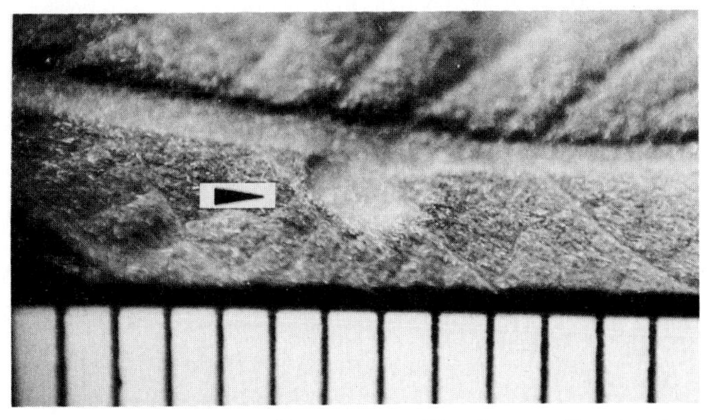

Photo 21: Young apple gall (arrow) on willow leaves collected in late April. At this time the gall contains a sawfly egg. View is of underside of leaf. Scale in mm.

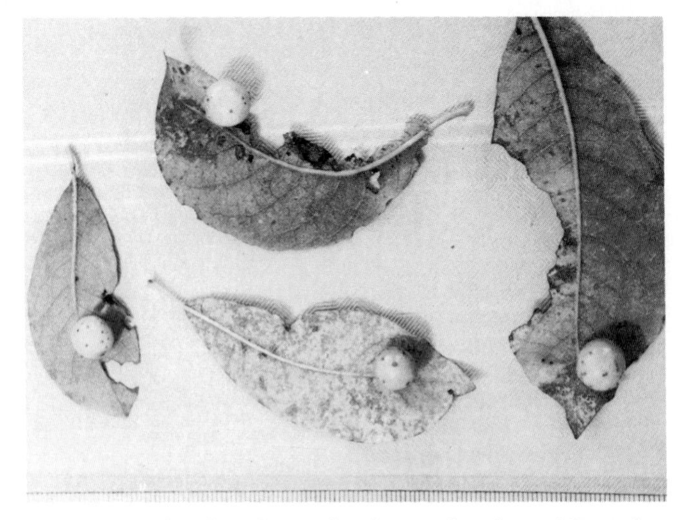

Photo 22: Apple galls on dying willow leaves collected in mid-September. View is of underside of leaf. Scale in mm.

This is considered a relatively primitive insect-caused gall, both because the insect and plant are primitive and because the gall itself is fairly simple in design. For example, no layer of thick-walled cells is formed. Instead the wall is made up entirely of unspecialized callus-like cells. Willow apple galls are often invaded by weevils (snout beetles) and by other predators and parasites.

For more information on the Willow apple gall see:

Caltagirone, L. E. "Notes on the Biology, Parasites, and Inquilines of *Pontania pacifica* (Hymenoptera: Tenthredinidae), Leaf-gall Incitant on *Salix lasiolepis.*" *Annals of the Entomological Society of America* 57, no. 3 (1964): 279-291.

Smith, E. L. "Biology of Gall-making Nematine Sawflies in the California Region." *Annals of the Entomological Society of America* 63, no. 1 (1970): 36-51.

Other Galls

Host: *Salix* spp.

If you find lumpy swellings on willow twigs, they are most likely caused by sawflies in the genus *Euura*. Depending on the species of sawfly, the swelling may involve one side of the stem (and look like drops of wax on a candle) or its entire girth. There are also a few cecidomyiid (midge) galls on the stems and leaves of willows in our area.

Small (2 to 3 mm in diameter) red or yellow blisters on willow leaves are caused by eriophyoid mites, perhaps *Eriophyes oenigma* Walsh. The blisters or capsules (photo 23) protrude mostly from the upper leaf surface but also slightly from the lower surface. These are single-chambered galls filled with several mites in the summer, and the gall chamber is incompletely partitioned by wall growths (photo 24). Several layers of enriched food cells line the chamber. The galls are first seen in early May. When the second flush of leaves comes on in midsummer, they are also galled. Little is known about the biology of the willow-gall mites.

Photo 23: Blister galls on willow leaf collected in late July. Scale in mm.

Photo 24: Cutaway view of blister galls on willow leaves collected in late May. The tops of the galls have been removed. The gall chamber is incompletely partitioned (arrow).

Pocket Leaf Gall

Host: *Populus trichocarpa* T. and G.

Gall Former: Aphid. *Thecabius populi-monilis* (Riley) Palmer

This is a common gall on black cottonwoods. In spring a single black aphid stem mother produces a red pocket, 5 to 15 mm long and 2 mm wide, that bulges up from the upper leaf surface. The gall lies between the midvein and leaf edge and runs parallel to the midvein. The underside of the leaf shows only the gall's slit opening. Usually only a single gall occurs on a leaf. The galls have a single chamber and, depending on the time of spring, contain one to several aphids.

As the stem mother produces young in midsummer, they move out of the spring galls and travel to the new flush of summer leaves, where they produce many of their own summer galls (photo 25). The summer galls look very much like the spring galls, except that they are usually wider (5 to 7 mm) and are arranged end-to-end for the length of the leaf. They look like sausage links and have a single chamber each. Several rows of the summer galls occur on a single leaf, with the result that the leaf is often curled. The summer foliage is often very heavily galled by these aphids.

Photo 25: Pocket leaf galls on black cottonwood collected in mid-August. These galls are formed by summer-generation aphids on the late flush of cottonwood foliage. Scale in mm.

The galls are fairly simple in construction and design. The wall of both spring and summer galls contains no layer of thick-walled cells and no food cells line the aphid chamber. Unlike most free-living aphids that feed on phloem sap, these and other gall-forming aphids feed for part of their lives on the contents of non-vascular cells. Unlike other species of gall-forming aphids on cottonwood, *T. populi-monilis* probably completes its entire life cycle on black cottonwood.

Leaf-Petiole Gall

Host: *Populus trichocarpa* T. and G.
Gall Former: Aphid. *Pemphigus* spp.

Throughout the western states, several species of *Pemphigus* aphids form galls on cottonwoods at the leaf-petiole junction. Very young attacked leaves show a slight thickened and twisted petiole (leaf stem) and leaf blade (photo 26). This early twisting and swelling probably takes place in a matter of a few hours. If you untwist the gall while it is young, the long slit on the lower gall surface opens and you see a single, fairly large black stem-mother aphid in the single-chambered gall. The stem mother controls gall growth until the gall reaches its full size. Young galls are reddish green, while older ones are red to dark green and eventually brown. A full-grown gall is 12 to 15 mm in diameter.

By the middle of May, the stem mother has started to produce young which will eventually fill the gall chamber. Small balls of wax-covered liquid also begin to accumulate in the chamber. These are excess plant saps that have been taken up and then excreted by the aphids. Most of the aphids leave the galls in late summer. Empty dead galls are common in the autumn leaf litter under any black cottonwood. We are not sure what happens to

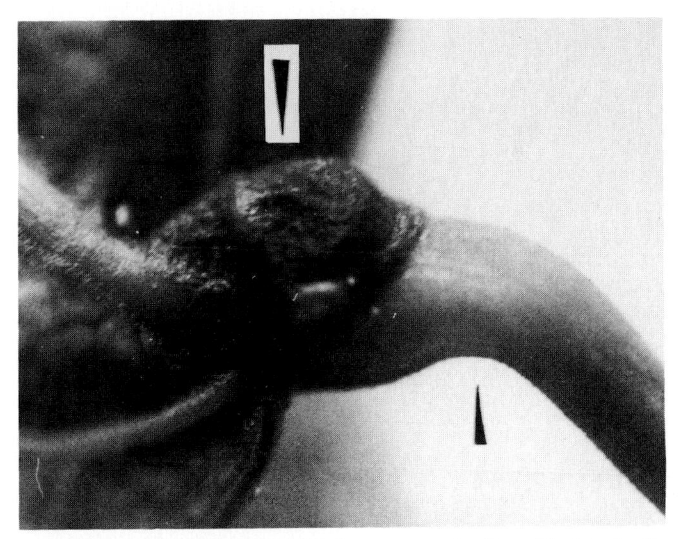

Photo 26: Leaf-petiole gall on black cottonwood collected in late April. The petiole (lower arrow) is bent and thick, and the leaf blade is curled around the petiole (upper arrow). This young gall contained a single stem-mother aphid.

the aphids in late summer; they probably migrate to the roots of grasses, beets, lettuce, or crucifers and then return to cottonwoods in the early spring.

Like the pocket leaf gall, the structure of the leaf-petiole gall is fairly simple. As in other aphid galls, the gall cavity is not lined by enriched food cells, nor is there a layer of thick-walled cells present. Little is known about the predators and parasites of poplar gall aphids. Recently the competition between stem mothers for appropriate-sized petioles has been studied.

For more information on Cottonwood aphid galls see:

Palmer, M. A. *Aphids of the Rocky Mountain Region.* Thomas Say Foundation vol. 5. Denver: Hirschfeld Press, 1952.

Whitham, T. G. "Habitat Selection by *Pemphigus* Aphids in Response to Resource Limitation and Competition." *Ecology* 59, no. 6 (1978): 1164-1176.

Big Bud

Host: *Populus trichocarpa* T. and G.
Gall Former: Gall Midge (Cecidomyiid). Undescribed.

These galls (photo 27) are most easily spotted during the fall and winter when the leaves are off the trees. The gall is an enlarged terminal or side bud and will stay attached to the plant for some time, even after the gall has died and turned brown and brittle. In fact, a dead big bud from the previous year can frequently be found on the same twig with a current year's big bud.

Attacked buds begin to swell in early May and are often covered with a yellow sticky resin. The bud is enlarged because the bud scales and stipules are thicker than normal. Nestled between the thick scales, you should be able to find several very small orange larvae that are often bathed in yellow plant resin. These are the midge larvae. They use the spaces between scales as their larval chambers.

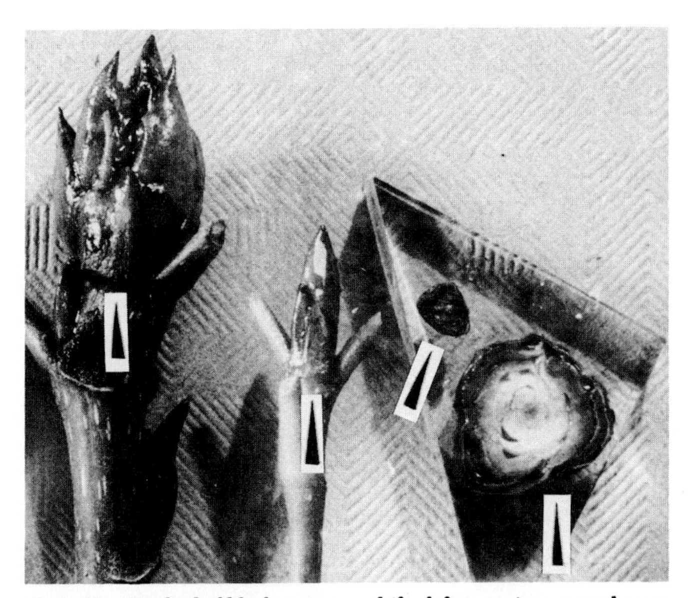

Photo 27: Big bud of black cottonwood (far left arrow) compared to an ungalled bud (center left arrow) collected in mid-August. Compare section through a big bud (far right arrow) to section through an ungalled bud (center right arrow).

In August, the fully formed galled buds are green and about 2.5 cm long and 1.5 cm wide. All bud parts are enlarged. The outer scales are green, and the inner ones are yellow to cream-colored. By mid-October, the 5 to 15 larvae in each bud are white (because of stored fat) and have a prominent sternal spatula. Most larvae occur near the center of the big bud and are often awash in the sweet-smelling resin. The bud tissue near a larva is sometimes brown. The larvae probably feed by scraping the cells on the surface of the bud parts. Feeding may cause the browning.

Other Galls

Host: *Populus trichocarpa* T. and G.

There is another aphid-caused gall on leaves of black cottonwood that is less common than the types already described. This pocket gall occurs along the leaf midvein. Unlike the *T. populimonilis* gall, this gall bulges out from the leaf's *lower* side, and the slit opening is seen on the leaf's *upper* side. The gall is probably caused by *Pemphigus balsamiferae* or *P. populivenae* (some entomologists argue that these two are the same species). The gall is red, 4 to 12 mm long and 2 to 3 mm wide. It is formed in April by a black stem-mother aphid, and by late spring it is filled with her young. The gall is vacated in the summer. Most likely, the aphids move to the roots of grasses, beets, or crucifers.

The leaf gall caused by the fungus *Taphrina aurea* (Pers.) Fr. appears as a light-green or yellow oval spot about 12 mm long and 7 mm wide (photo 28). The oval is raised slightly on the upper leaf surface, with a corresponding depression about 2 mm deep on the leaf's lower surface. This is considered a gall because the yellow tissue is thicker than the neighboring green leaf-blade tissue. The spots begin to turn black in the late summer. Several of these galls may occur on a single leaf.

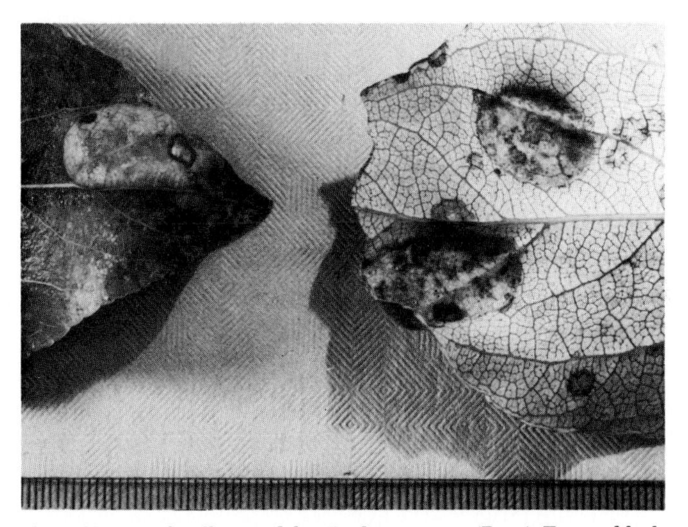

Photo 28: Leaf gall caused by *Taphrina aurea* (Pers.) Fr. on black cottonwood collected in early August. View is of leaf's upper surface on left and lower surface on right. Scale in mm.

A fairly common gall on black cottonwood caused by eriophyoid mites, *Eriophyes parapopuli* Keifer, occurs at the base of buds along the stem. When branches are bare in the autumn, the galls are woody, brown, cauliflower-like, and rarely more than 2.5 cm in diameter. Several occur along a single stem. The mites live in the cracks and fissures of the gall.

Long-horned beetles, (*Saperda populnea* L. or a closely related species) cause spherical or spindle-shaped stem galls (2 cm in diameter) on black cottonwood and on willows. Infestations are often heavy; the shape of the tree may be somewhat disfigured as a result. Adult females lay eggs in late spring in holes that they chew in the bark. When the egg hatches, the larva mines the cambium and then turns into the wood where it excavates a gall chamber. Growth from the wounded cambium causes the swelling. One or two adults emerge from each gall after one to two years of larval development. The adults feed on poplar leaves.

Mossy Gall (Bedeguar Gall)

Host: *Rosa eglanteria* L.
Gall Former: Gall Wasp (Cynipid). *Diplolepis rosae* L.

In the Pacific Northwest this gall (photo 29) occurs on the introduced wild eglantine rose (also known as sweetbrier). It is a distinctive and very common gall. Plants, especially those in the sun, are often thick with galls. Throughout its development, the gall is covered by a shag that looks very much like moss. This is actually a heavy growth of long plant hairs. The gall has a number of chambers, each holding a single gall-wasp larva. Several white grubs can be found in these galls in summer or fall. Some of these are larvae of the gall-former; others are larval parasites.

The gall first appears in mid-to-late spring as a small red or green mossy patch on a leaflet, petiole, young stem or young flower bud. The galls continue to grow so that at maturity in midsummer they range from 2.5 to 5.0 cm in diameter. Each larval chamber is lined by enriched food cells. For most of the summer, the hairs on the galls are green, but in October, they begin to turn yellow-brown and are brown through the winter. The gall becomes woody in late summer; although larvae over-winter in the gall, they apparently do not feed during the winter,

Photo 29: Mossy gall on *Rosa eglanteria* L. collected in mid-August. Scale in mm.

because the wall of their cavity becomes dry and hard in late fall. Galls stay on the plant well into the following spring, if not longer; often old galls that have rotted or that have been pecked apart by birds remained attached to the plant for a year.

Adult wasps emerge from the galls in late winter to early spring. If you are in a field of sweetbriers at emergence, you will see the small (4 mm long) reddish-brown females, which are clumsy fliers, at the tip of an unopened or just-opened bud. If conditions are right, each wasp will maneuver her ovipositor (a hollow flexible thread-like tube attached to her hind end) down into the bud to deposit her eggs. The insects are easy to rear from the galls. A few galls brought indoors in late winter or early spring will provide you with plenty of adults. If you provide these adults with a few clipped pieces of sweetbrier stems you can watch them lay their eggs at your leisure. Mossy rose galls to not occur on cultivated hybrid tea, multiflora, or floribunda roses.

A very similar mossy gall occurs on the twig tips of the wild Nootka rose (*Rosa nutkana* Presl.). It is caused by the gall wasp (cynipid) *Diplolepis bassetti* (Beut.) Weld.

Tip Gall

Host: *Rosa nutkana* Presl.

Gall Former: Gall Wasp (Cynipid). Probably *Diplolepis oregonensis* (Beut.) Weld.

This gall (photo 30) looks like a rose hip. It is spindle-shaped and occurs on the twig tips of the wild Nootka rose. It has a number of chambers and is 12 to 15 mm long and 7 mm wide at maturity. Each chamber contains a single larva. Through the summer the gall is green, but in winter and the following spring, it turns a very light brown. Several small black gall wasps emerge from the gall in early spring. Although known in the Pacific Northwest for many years, this gall has not received much scientific attention.

For more information on Rose gall wasps see:

Weld, L. H. *Cynipid Galls of the Pacific Slope (Hymenoptera, Cynipoidea): An Aid to their Identification.* Ann Arbor: privately printed, 1957.

Photo 30: Tip galls on *Rosa nutkana* Presl. collected in mid-August. Scale in mm.

Stem Gall

Host: *Rubus parviflorus* Nutt.
Gall Former: Gall Wasp (Cynipid). *Diastrophus kincaidii* Gill.

These galls (photo 31) are easiest to find in the late fall or winter, when the stems are leafless. In October, the galls are green or brown, anywhere from 2.5 to 4.0 cm long, 2 to 2.5 cm wide, and contain many white larvae, each in a larval chamber. The galls usually occur slightly higher than halfway up the stem, and they may stunt or bend the stem. The entire girth of the stem is swollen at the gall.

When young, the galls are green or yellow-green. You can first spot them in early summer not long after the adult gall wasp females have emerged from last year's galls and have laid their eggs in young stems.

For more information on the Thimbleberry stem gall see:

Wangberg, J. K. "*Biology of the Thimbleberry Gall Maker, Diastrophus kincaidii.*" *Pan Pacific Entomologist*, no. 1 (1975): 39-48.

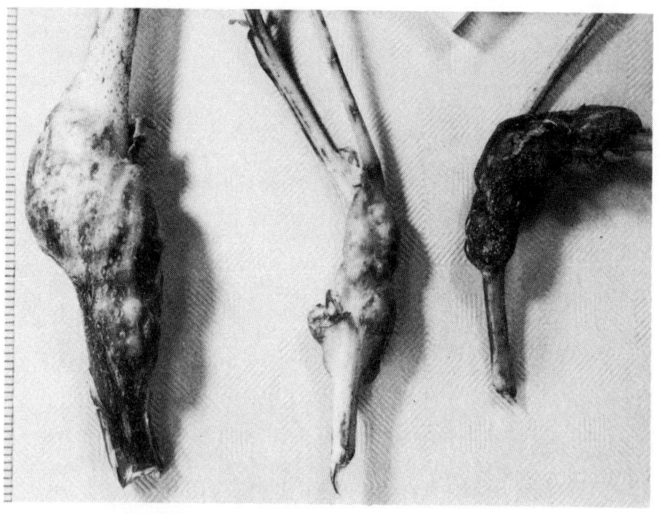

Photo 31: Stem galls on thimbleberry collected in mid-August. Scale in mm.

Stem Gall

Host: *Hypochaeris radicata L.*

Gall Former: Gall Wasp (Cynipid). *Aylax (Aulax) hypochaeridis* Kieff.

False dandelion is an introduced lawn and roadside weed that looks like a long-stemmed dandelion. The galls (photo 32) occur along the midsection of the flower stem, and the entire girth of the stem is swollen at the gall. Galls are slightly ribbed, green (when young) or yellow-brown (in August), and range in size from 2.5 to 6.0 cm long and 1.5 cm in diameter at maturity. The gall's many chambers each contain one gall wasp larva. Galls first become noticeable in late May to early June; at this time, they contain very small larvae embedded in the white spongy pith tissue. However, from then until fall, the larval chamber is lined by green, enriched food cells. The chambers are scattered along the length of the gall.

The wasp and its gall may help control the spread of the weed, but this needs to be studied. We do not know, for example, whether plants with galled stems produce few or poor-quality seeds. The structure of the gall has been studied in Europe, but otherwise little research has been done with this gall.

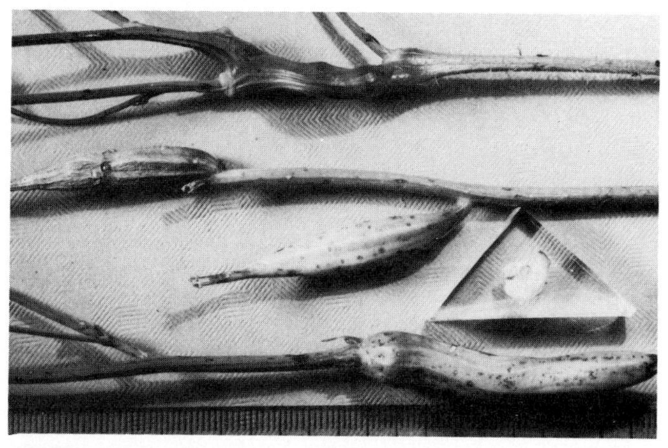

Photo 32: Stem galls on false dandelion collected in early August. Scale in mm.

Leaf Galls

Host: *Crataegus* spp.
Gall Former: Gall Midges (Cecidomyiids).

A variety of gall midges attack hawthorn leaves. In fact, the chances are good that any gall you find on wild or ornamental hawthorn is a midge (cecidomyiid) gall. Unfortunately, many of the midges on *Crataegus* are undescribed.

One of the more common galls (photo 33) in the Pacific Northwest occurs on Black hawthorn (*C. douglasii* Lindl.). The leaf folds up at the midrib, and the midrib and small portions of the adjoining blade swell, so that a long (4.5 cm) tube-like cavity forms down the length of the fold. The lips of the gall (the curled thickened leaf blade) are pressed together so that the larval chamber is closed. Numerous (10 to 20) larvae live communally in the single larval chamber. From mid-May through most of the summer, the larvae are very small. By late September, they are white and are large enough so that they almost completely fill the chamber. By October the leaves and galls are brown, although the gall may stay green for a while longer than the leaf blade. Most likely, the larvae or pupae overwinter in the fallen galls, and adult midges emerge and lay their eggs on young leaves in spring.

Photo 33: Leaf galls on hawthorn collected in early August. Galls are viewed from side. Ungalled leaf is shown at lower right. Two cross sections of gall are in upper right corner. Scale in mm.

Leaf Gall

Host: *Amelanchier alnifolia* Nutt.
Gall Former: Gall Midge (Cecidomyiid). Perhaps *Trishormomyia canadensis* Felt or a closely related species.

This pouch gall, with one larva in a single chamber, occurs on the leaves of serviceberry (photo 34). The bottom of the pouch protrudes from the lower side of the leaf and looks something like a flat-sided teardrop. The top of the pouch tapers to a short, narrow neck, the rim of which is raised slightly above the upper leaf surface and is surrounded by a bit of white fuzz. When completely removed from a leaf, the gall is 5 mm long, 2 mm across, and 1 to 2 mm deep. Galls may occur singly on a leaf or often in groups of 10 to 20 per leaf. When they occur in high densities, the leaf is curled, and its underside appears to be covered with cotton.

The structure of this gall is surprisingly complicated. Instead of enjoying a spacious chamber, the midge larva lives in a small tunnel that runs along the bottom of the pouch and up both sides. The tunnel becomes closed early in gall development and is lined by enriched cells. As the gall matures, however, the tissue lining the tunnel becomes less and less enriched.

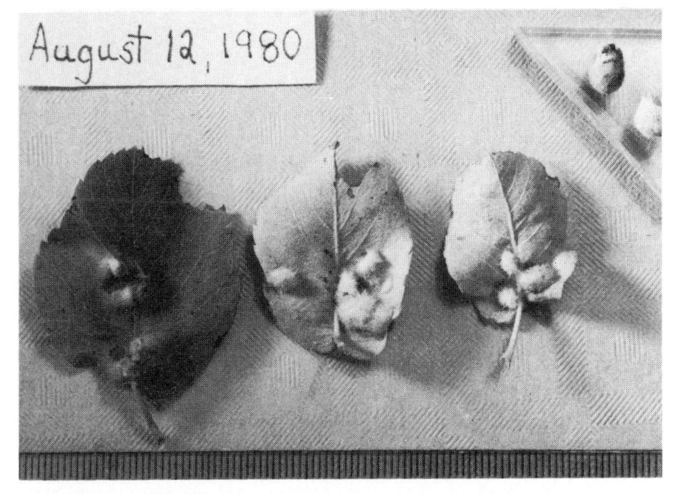

Photo 34: Leaf galls on serviceberry collected in mid-August. Upper leaf surface is seen at far left. Two leaves on right show the lower leaf surface. Two galls removed from leaf are in upper right corner. Scale in mm.

The chamber in the center of the gall extends up the neck to the rim; it remains uninhabited through the summer. In October, however, the gall begins to dry, and the constriction between the tunnel and central chamber opens, so that the larva has free run of the entire gall. We have no idea what purpose the empty central chamber plays in early gall development. Interestingly, several galls, especially those formed by wasps, have similar empty chambers.

As it grows, the gall develops stiff walls that become more and more difficult to cut through. The stiffness is due to a thick layer of plant cells that develop reinforced walls. By late September, when serviceberry leaves begin to fall, the gall larva is still growing. Presumably the midge larva or pupa overwinters in the fallen gall, and the adult emerges in early spring.

Terminal Leaf Gall

Host: *Symphoricarpos albus* Blake.
Gall Former: Sawfly (Tenthredinid). Undescribed.

This gall (photo 35) appears in early spring. The young leaf is completely disrupted so that the gall looks more like a bud than a leaf. Galls often occur in pairs at the branch tip, and the pair may share a common wall. The galls are capsule-like, green, generally spherical to egg-shaped, and 12 to 20 mm in diameter at maturity. The gall walls are fairly thick and enclose a larval chamber in which a single, caterpillar-like larva lives and defecates. The walls show no complex layered organization, and the chamber is lined by an enriched callus. The walls frequently are riddled with holes. These may be caused by predators; more likely, they are made by the gall former as windows through which fecal pellets can be pushed. The sawfly probably overwinters in the leaf litter and emerges as an adult in the early spring. Empty, dry galls often remain on the plant for a year.

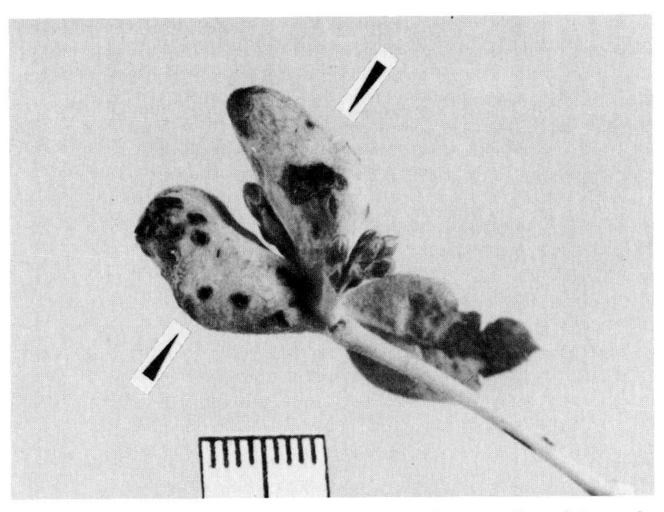

Photo 35: Terminal galls (arrows) on snowberry collected in early August. Leaves have been removed to show galls more clearly. Holes in gall are common. Scale in mm.

Leaf Roll Gall

Host: *Symphoricarpos albus* Blake.
Gall Former: Gall Midge (Cecidomyiid). Perhaps *Thomasia californica* Felt, but probably undescribed.

This is an easily spotted and very common gall on snowberry, generally occurring on leaves near the branch tip (photo 36). Gall midge larvae cause the leaf edge to roll up and in toward the midvein. If both edges of the leaf are rolled, the leaf looks like a scroll. The rolled leaf blade becomes thick. A long, tube-like tunnel runs through the center, down the length (2.5 cm) of the roll, forming a chamber in which 5 to 15 larvae live and feed. Thus, this is a multilarval, single-chambered gall.

The galls first appear in early May and reach full size by midsummer. When young, the gall is green to yellow-green, but in late July it begins to yellow. Because of their size, larvae are difficult to see in young galls, but by July, the insects are white and fairly large (1 to 2 mm long). In October, when all of the other leaves have fallen, many of the galled leaves remain on the plant. At this time, galls uncurl slightly, and the gall walls begin to dry, harden, and turn brown; the larvae continue to live in the gall. The insects probably overwinter as larvae in leaf litter on the ground.

Photo 36: Leaf roll gall on snowberry collected in mid-July. Upper leaf viewed from underside. Lower leaf viewed from upper surface. Scale in mm.

Leaflet Gall

Host: *Rhus diversiloba* T. and G.

Gall Former: Gall Mite (Eriophyoid). *Aculops toxicophagus* Ewing.

Those of you who suffer from poison oak dermatitis can take some small comfort in the fact that the plant has its problems, too! Eriophyoid mites form red galls on its leaflets (photo 37). These galls protrude slightly from the upper leaf surface; they are round, 1 to 2 mm in diameter, and are often found in clusters. Heavy infestations cause galled leaves to fold and twist. The galls develop in late spring through late summer. They are actually small pouches or out-pocketings of the leaf surface. Several mites live in the chamber of each gall.

A similar (if not the same) mite also galls poison oak flowers (photo 38). Attacked flowers are red, hairy, and swollen. They remain unopened, and their petals are often stunted, fused, or swollen. Gall mites applied in a spray might be used to slow the spread of this undesirable plant. Tests to determine the amount of damage caused by the leaf and flower mites, however, have not been performed—no one is very interested in tangling with the plant!

Photo 37: A patch of leaflet galls (arrow) on poison oak collected in late May. View is of upper leaflet surface.

Photo 38: Flower galls on poison oak collected in early July. Scale in mm.

Big Bud

Host: *Corylus avellana* L. and *C. cornuta* Marsh.

Gall Former: Gall Mites (Eriophyoids). *Phytoptus avellanae* (Nal.) Keifer and *Cecidophyopsis vermiformis* (Nal.) Keifer.

These mites gall buds of both cultivated and wild filberts. Infested buds begin to swell in spring. Some reach full size and fall off the trees in midsummer (summer big buds), while others swell more slowly, reach full size in autumn, and fall off in spring (spring big buds) (photo 39). Spring big buds, the more common of the two types, are spherical, 12 to 15 mm in diameter, and look something like a piece of popcorn. When young, they are green; as they grow and burst open, they become yellow-green, red, and finally dark brown.

When you open a mature bud, you see hundreds of very small mites crawling over the surfaces of the swollen bud parts. Most of the mites are concentrated in the outer whorls, but they also crawl to the center of the bud. If you look closely, you will see that much of the surface of the bud parts is carpeted with many small, fleshy, finger-like projections. The mites crawl on and between these projections and feed on the surface cells (photo 7).

In late winter, mites begin to migrate, either by wind or by walking, to young buds at the stem's tip. A second migration occurs in midsummer when mites leave the dying summer big

Photo 39: Spring big bud of filbert collected in mid-July.

buds. Both species of mites live in both types of galls. *P. avellanae, however, is more prevalent in spring big buds, and C. vermiformis* is more common in summer big buds. We are not sure how the two species interact in the galls.

The big bud problem is usually not serious enough to cause concern. Resistant varieties of filberts are available, and breeding programs at Oregon State University continue to select for resistance. Recently, however, plant pathologists have begun to suspect a link between big bud and filbert blight disease. If a link exists, then big bud infestation will be viewed as a much more serious problem.

For more information on Filbert big bud galls see:

Jeppson, L. R., H. H. Keifer, and E. W. Baker. *Mites Injurious to Economic Plants.* Berkeley: University of California Press, 1975.

Leaf Curl Gall

Host: *Arctostaphylos uva ursi* (L.) Spreng and other *Arctostaphylos* species.
Gall Former: Aphid. *Tamalia coweni* Ckll.

Bearberry, a common ornamental ground cover, often has several curled, cream-colored or pink leaves at the end of its stems (photo 40). Stems with several attacked leaves look like short flower stalks lying on the ground. The leaf margins fold under the lower leaf surface, and the curled part of the leaf (1 to 1.5 cm long) is thicker than the uncurled leaf blade.

A single black stem-mother aphid forms each single-chambered gall in early spring; by late spring, she has begun to fill the chamber with her young. In summer, you should find several aphids in each gall. By October, the aphids have left the gall, so that all you see when you open a gall are the several skins left behind by the aphids. We are not sure where the aphids overwinter. Probably the easiest way to eliminate or minimize the damage to your ground cover is to remove and burn the galls in the spring or summer.

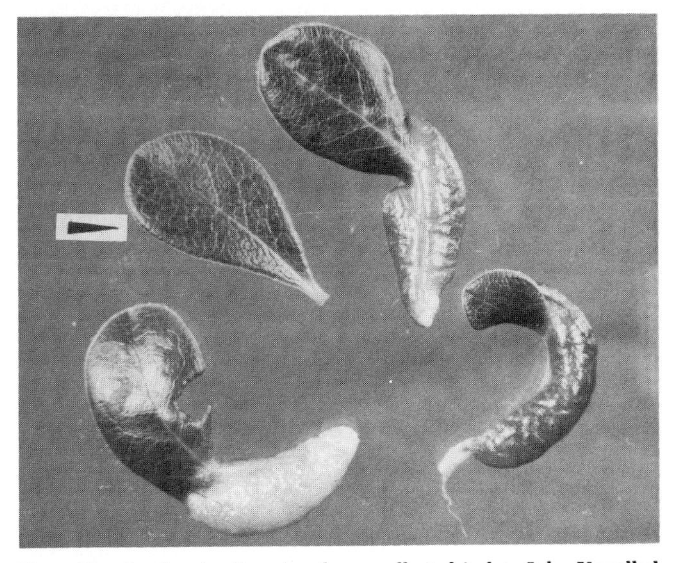

Photo 40: Leaf curl gall on bearberry collected in late July. Ungalled leaf at arrow.

Leaf Gall

Host: *Tilia* spp.

Gall Former: Gall Mite (Eriophyoid). Perhaps *Phytocoptella (Eriophyes) abnormis* (Garman) Keifer.

Many of the streets in western Washington and Oregon towns are planted with introduced basswood (lime, as the British would say). In mid-April, some of the leaves will develop small green blisters on the upper surface that, as the summer progresses, become long and pointed, sometimes curled at the tip (photo 41). The mature, single-chambered red gall looks like a narrow pointed tube sticking up from the leaf surface.

One to several galls, each about 7 mm long, may occur on a leaf. Each gall is formed by a single female. In mid-May, she begins to produce offspring that fill the long, cylindrical, hairy gall chamber. The overwintering life stage is unknown, but we suspect that it is an egg or hardy female (deutogyne). Damage to the tree is usually negligible.

Photo 41: Leaf gall on linden collected in late May. Gall arises from leaf's upper surface.

Leaflet and Bud Galls

Host: *Fraxinus latifolia* Benth.
Gall Former: Gall Mites (Eriophyoids).

In western Oregon and Washington, the leaflets of ash are galled by eriophyoid mites (photo 42). The leaflets bear one to several single-chambered blisters, 1 to 2 mm in diameter; each gall contains only a single orange mite until late June, when the gall chamber becomes filled with her white offspring. If you slice one of these galls open, you see that the chamber is partially divided by gall wall partitions.

Eriophyoids (most likely *Eriophyes fraxiniflorae*) also live in and probably cause a witches' broom gall of ash buds in the Pacific Northwest. Instead of elongating into normal stems, the buds remain short and produce many small, tightly packed branchlets with deformed leaves. The mites live between and at the base of the branchlets.

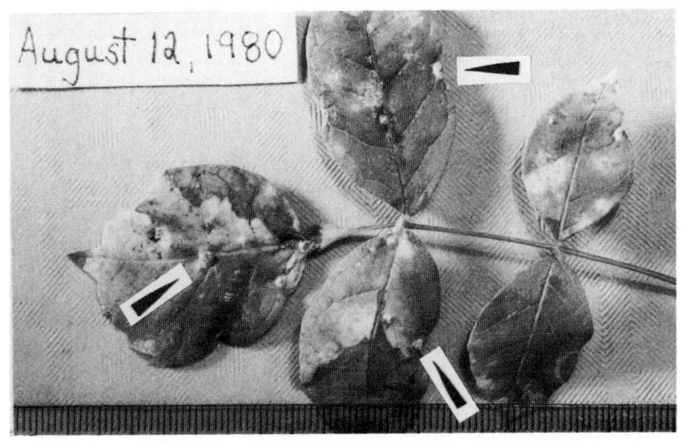

Photo 42: Galls on ash leaflets (arrows) collected in mid-August. Scale in mm.

Leaf Curl Gall

Host: *Ulmus americana* L.

Gall Former: Aphid. *Eriosoma americanum* (Riley) Palmer.

As long as American elms survive in the western valleys of Oregon and Washington, they will be attacked by gall aphids, most commonly by *E. americanum*. An attacked leaf rolls under from the margin toward the midvein (photo 43). The rolled part of the blade becomes thick and yellow. The gall is usually 2.5 to 5 cm long. Several aphids live in the curl, while others feed at the entrance to the curl. The single-chambered gall is initially formed by a stem mother in spring, and by late summer it is filled with her progeny and several wax-colored balls of plant sap. According to some reports, the aphids migrate from the elm galls to the roots, stems, and leaves of serviceberry in late summer, but no galls are formed on this secondary host. The stem mother reappears on elm in early spring.

Another species, *E. lanigerum*, the woolly apple aphid, causes leaf cluster galls on elm. To complete their life cycles, the aphids migrate from elms to the branches, trunks, and roots of apple trees. Apple roots swell where the aphids feed, and these root galls may seriously decrease orchard yields. This insect is of economic importance not because of the leaf galls it causes on elms, but because of the root galls it causes on apples. Neither species of *Eriosoma* has been implicated in the spread of Dutch elm disease.

For more information on Elm leaf curl galls see:

Metcalf, C. L., W. P. Chalk, and R. L. Metcalf. *Destructive and Useful Insects*. 4th edition. New York: McGraw-Hill, 1962.

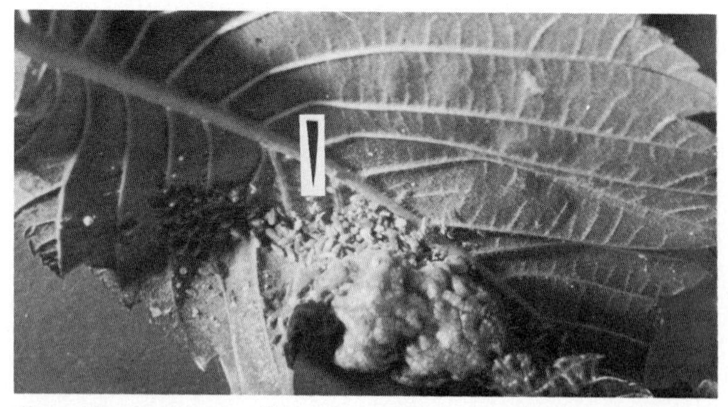

Photo 43: Leaf curl gall on elm collected in early July. View is of underside of leaf. Aphids (arrow) are visible.

Bead Gall

Host: *Alnus rubra* Bong.
Gall Former: Gall Mite (Eriophyoid). Probably *Phytoptus laevis* Nal.

Alder leaves are often peppered with these small bead galls (photo 44). When they first appear in mid-April, the galls are about 1 mm in diameter and are red or green, turning yellow or pink as they grow. At maturity, the single-chambered gall is about 2 mm in diameter and protrudes through both surfaces of the leaf. The top of the gall is flat, while the bottom is round or pointed.

Each gall is formed by a single, orange, overwintering female (deutogyne). She is the only mite in the large gall chamber until mid-July, when she begins to lay eggs; her young (up to 400) eventually fill the chamber. The mites feed on the contents of the large enriched cells that line and protrude slightly into the gall chamber. Many of the galls are brown by late summer. Although studied in the U.S.S.R., more information is needed on the life cycle of this mite in the U.S.

For more information on Alder bead galls see:

Shevtchenko, V. G. "The Life-history of Alder Gall Mite, *Eriophyes laevis* (Nalepa)." Summary in English, *Revue d' entomologie de l'URSS* 36, no. 3 (1957): 598-618.

Photo 44: Bead galls on alder collected in mid-July. The leaf's upper surface is seen at left, the lower surface at right. Scale in mm.

Bladder Gall

Host: *Acer saccharinum* L. and *A. rubrum* L.
Gall Former: Gall Mite (Eriophyoid). *Vasates quadripedes* (Shimer)

This mite attacks both silver and red maples. Adult female mites overwinter in bark cracks; in early spring, they move to the underside of young leaves. Feeding causes a yellow or red pouch or bladder on the leaf's upper surface (photo 45). The female and her progeny are found in the bladder. During the growing season, the progeny leave the gall through a small portal on its underside and establish new bladder galls on the same or other leaves. At maturity the galls are 2 to 7 mm in diameter and are frequently found in dense clusters. Usually the tree is not significantly damaged.

For more information on Maple bladder galls see:

Keifer, H. H. "A Review of North American Economic Eriophyid Mites." *Journal of Economic Entomology* 39, no. 5 (1946): 563-570.

Photo 45: Bladder galls on maple collected in mid-July.

Erineum

Host: *Acer* spp.
Gall Former: Gall Mite (Eriophyoid). *Eriophyes* spp.

There are scattered observations of bright red erinea (photo 46) on both native and ornamental maples in western Oregon and Washington. These need to be confirmed and the identity of the mites determined. Several species of eriophyoid mites cause red, velvety erinea on either the upper or lower surface of maple leaves in the U.S. Although these galls, because of their color, are very noticeable, usually the damage to the plant is not severe.

The mites overwinter as adults in bark crevices and move to young leaves in early spring. The dense patches of hairs that result from mite attack may be scattered over the leaf surface as circular or oval spots, concentrated at the leaf end, or fused to form large, irregularly shaped patches.

For more information on Maple erineum galls see:

Hodgkiss, H. E. *The Eriophyidae of New York. II. The Maple Mites.* New York Agricultural Experiment Station Technical Bulletin 163 (1930).

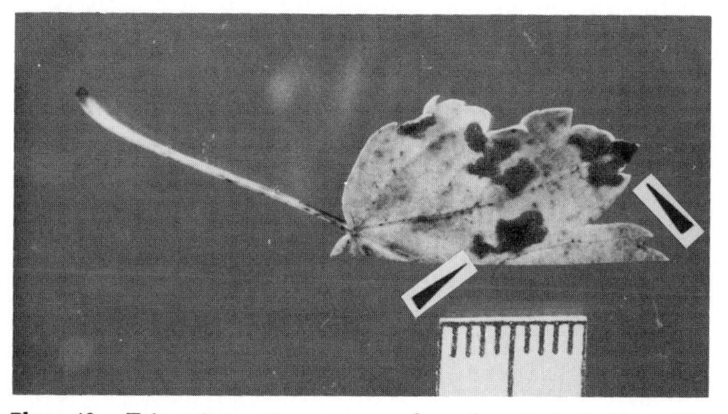

Photo 46: Erinea (arrows) on upper surface of maple leaf collected in mid-July. Scale in mm.

Pod Gall

Host: *Gleditsia* spp.

Gall Former: Gall Midge (Cecidomyiid). *Dasyneura gleditschiae* O.S.

First described from Rhode Island in 1866, this gall former is now found throughout North America wherever honey locusts are grown. The thornless locust varieties, such as Shademaster, Imperial, and Moraine are especially susceptible to attack.

The adult midge appears in April and deposits very small yellow eggs in young leaflets. Usually several neighboring leaflets on a single leaf are attacked. The larvae hatch in about 2 days and begin to feed immediately. In response to this feeding, the leaflets become globular and form pod-like capsules in May through June. Each pod is 4 to 8 mm in diameter and contains one to several white midges that are 6 mm long at maturity. Several generations may occur during the summer. The midge goes through its pupal stage in the gall. Galled leaflets dry up and are shed prematurely; repeated defoliation may cause twig death.

For more information on Honey Locust pod galls see:

Schread, J. C. *Pod Gall of Honey Locust.* Connecticut Agricultural Experiment Station, New Haven, Circular 206 (1959).

Erineum

Host: *Juglans regia* L.

Gall Former: Gall Mite (Eriophyoid). *Eriophyes erineus* (Nal.) Also known as *E. tristriatus erineus* Nal.

This mite, sometimes called the walnut blister mite, forms dense hairy patches (erinea) on the underside of English (Persian) walnut leaflets (photo 47). The erinea are yellow and velvety. They usually occur between the leaflet's secondary veins, and the veins bordering the erinea are swollen. A heavily infested leaflet may show several abutting erinea. The upper leaflet surface is puckered and yellow above an erineum.

The mites live at the base of the hairs during the early summer. In late summer and early fall, females move from the erinea to vegetative buds, where they overwinter. The following spring the mites attack the young leaflets. Damage to the tree is usually not severe.

For more information on Walnut erineum galls see:

Michelbacher, A. E. and J. C. Ortega. *A Technical Study of Insects and Related Pests Attacking Walnuts.* California Agricultural Experiment Station Bulletin 764 (1958).

Photo 47: Erinea (arrows) on walnut leaflets collected in mid-July. View is of underside of leaflet.

Erineum

Host: *Vitis vinifera* L.
Gall Former: Gall Mite (Eriophyoid). *Colomerus vitis* (Pgst.) Also known as *Eriophyes vitis* (Pgst.)

In addition to causing leaf erinea, strains of this mite are believed to also cause bud death and leaf curling. The erinea, which occur on the leaf's undersurface, are dense patches of single-celled hairs. When young, the white or yellow hairs are filled with grains of fat, but as they age, they turn brown with tannic material. The leaf structure beneath the hairs is altered by the mites in such a way that disease-causing fungi can more easily enter the plant through erinea than through the healthy leaf surface. Thus the gall is important both because of the damage it causes and the damage it allows. American varieties of grapes are more resistant to the mite than are European varieties.

For more information on Grape erineum galls see:

Slepyan, E. I. et al. "The Gall of the Mite *Eriophyes vitis* Pgst. (Acarina, Eriophyidae)." *Entomological Review* 48, no. 1 (1969): 67-74.

Leaf Blister

Host: *Pyrus* spp. (Cultivated pears). Also found on apple, mountain ash, cotoneaster, and serviceberry.

Gall Former: Gall Mite (Eriophyoid). *Phytoptus pyri* Pgst. (*Eriophyes pyri* Pgst.)

Rows of brown leaf blisters along the midvein of pear and apple leaves are often caused by these gall mites. The mites overwinter in an inactive state in the buds and bark crevices of the host. In March, they begin to feed and lay eggs on young leaves in swelling buds. The mites feed on the undersurface of the leaves and cause a surface swelling in localized spots. As the swollen cells of the young blisters die and rupture, the mites move into and feed in the leaf.

At maturity, the blisters are 1 to 2 mm in diameter and often abut one another. On the upper leaf surface the blisters are yellow-green when young, then red, and finally brown. On the leaf's under surface, the blisters are crusted or corky.

These mites may also feed on the surface of fruits, and heavy infestations may cause premature fruit drop.

For more information on Pear leaf blister galls see:

Jeppson, L. R., H. H. Keifer, and E. W. Baker. *Mites Injurious to Economic Plants*. Berkeley: University of California Press, 1975.

Parrott, P. J., H. E. Hodgkiss, and W. J. Schoene. *The Apple and Pear Mites*. New York Agricultural Experiment Station Bulletin 283 (1906): 281-318.

Needle Gall

Host: *Pseudotsuga menziesii* (Mirbel) Franco.
Gall Former: Gall Midge (Cecidomyiid). *Contarinia pseudotsugae* Condrashoff.

This gall midge is sometimes a serious pest in Christmas tree plantations of Douglas-fir. Adults emerge in the spring at bud burst, and the females lay eggs on young needles. The larvae feed internally about halfway along the length of the needles; they cause the midportion of the needle to swell slightly and turn yellow or purple (photo 48). Several galls may occur on a needle. In fall the larvae drop to the ground, and they pupate in early spring. When heavily attacked, a tree may lose many of its needles and experience some twig dieback. Intensity of attack varies from year to year.

Two other less common midges gall Douglas-fir needles. *Contarinia constricta* Condrashoff causes yellow galls towards the tip end of the needle (photo 49, lower arrow). *C. cuniculator* Condrashoff causes galls towards the basal end of the needles (photo 49, upper arrow).

For more information on Douglas-fir needle galls see:

Condrashoff, S. F. "Bionomics of Three Closely Related Species of *Contarinia* Rond. (Diptera: Cecidomyiidae) from Douglas-fir Needles." *Canadian Entomologist* 94, no. 4 (1962): 376-394.

Sander, G. H. *Douglas-fir Needle Midges.* Oregon State University Fact Sheet 164 (1969).

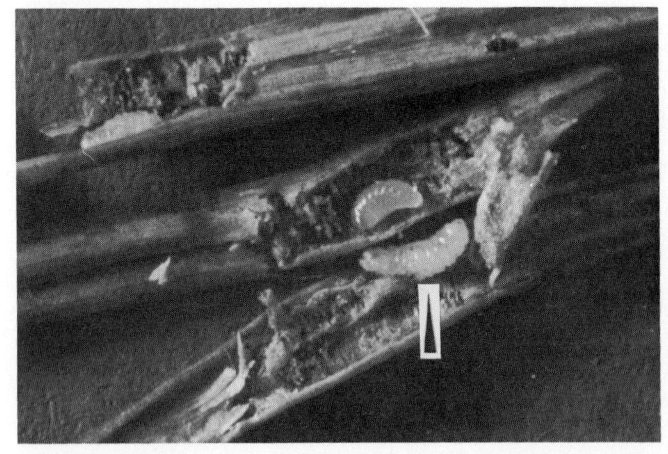

Photo 48: Opened midge (cecidomyiid) galls on Douglas-fir needles with larvae (arrow) exposed. Galls were collected in late June.

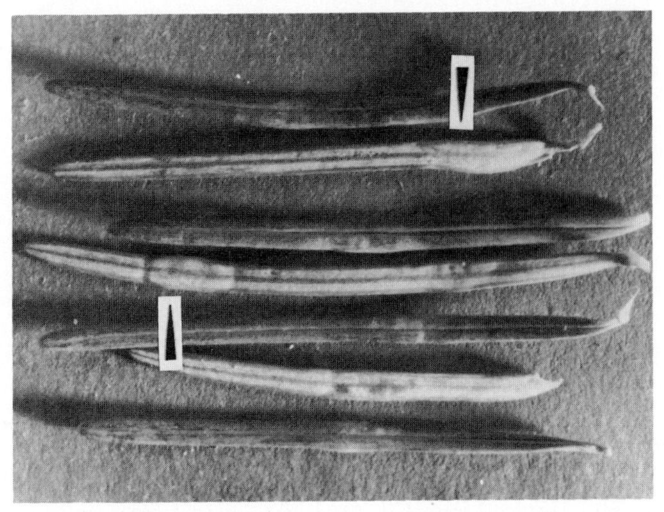

Photo 49: Midge (cecidomyiid) galls on Douglas-fir needles (arrows) collected in late June.

Anasas (Pineapple) Gall

Host: *Picea engelmannii* Parry, *P. sitchensis* (Bong.) Carr, and several other *Picea* species.
Gall Former: Adelgid. *Adelges coolyei* Gill.

If you have Engelmann or Sitka spruce in your yard, chances are that you know this gall very well. It is caused by the Cooley spruce adelgid. Adelgids are close relatives of the aphids. When young, the gall (photo 50) looks very much like a small green pineapple or pine cone on the spruce branch, often at the branch tip. Galls are 2.5 to 5.0 cm long and 2.5 cm in diameter at maturity.

The gall is formed from the swollen and fused bases of several neighboring needles. Each gall has several closed chambers, one at the base of each swollen needle. One to twenty small adelgids live and feed in each chamber. The adelgids secrete and carry streamers of white wax on their backs. As the galls dry out in late summer, the gall chambers open and the adelgids migrate to Douglas-fir. Although no gall is formed on Douglas-fir, the clusters of adelgids on needles and twigs are very noticeable on this host plant because of the dense cottony secretion that covers the insects. In addition to switching hosts, these insects have a complicated, two-year life cycle with several different life stages.

Photo 50: Anasas gall on spruce (arrow) collected in early July. Scale in mm.

The formation of the anasas gall begins when a single stem-mother adelgid settles at the base of a bud and causes a small amount of swelling. She soon produces young that move to the base of nearby needles in the bud and continue the gall-forming process until the gall is mature.

The damage caused by this adelgid is usually not significant. On the other hand, the closely related *A. piceae* (Ratz.), commonly called the balsam woolly aphid, is a major forest pest of the true firs. These insects feed externally on the plant, and cause gouting (galling) of twigs as well as irregularly grained wood at feeding sites on the bole. When combined, these afflictions may kill a tree in a few years. *Adelges nusslini* (Borner), which causes needle distortion on several of the true firs in western Oregon and Washington, also develops on spruce. The damage is usually mild.

For more information on Spruce anasas (pineapple) galls see:

Furniss, R. L. and V. M. Carolin. *Western Forest Insect.* U. S. Forest Service Miscellaneous Publication no. 1339 (1980).

Pine Louse Gall

Host: *Picea engelmannii* Parry, *P. sitchensis* (Bong.) Carr, *Pinus monticola* Lamb.
Gall Former: Adelgid. *Pineus pinifoliae* (Fitch.).

This is occasionally a serious problem on ornamental spruces. The adelgids cause cone-shaped galls on the branch terminals. The galls resemble those caused by *Adelges cooleyi*. The chambers in the *Pineus* galls are distinct, however, because they are interconnecting. Each chamber usually holds only one or two insects. Spruces are the primary host. No galls are formed on the secondary host, mountain white pine, but the needles and stems of pines are often heavily damaged by the insect.

Another species of *Pineus*, *P. boycei* Annand, causes cone-shaped galls on Engelmann spruce. The gall chambers are interconnected, and each holds about 15 nymphs.

Leaf-Curling Aphids

Several aphids cause leaf curl or crinkling on various ornamentals (photo 51). Usually the attacked leaf swells (i.e., is galled) slightly. It may remain green or turn yellow or red, and it folds or crumples; in so doing, it encloses the colony of aphids. Following is a list of aphids that cause these symptoms on various plants.

HOST	APHID
Creek Dogwood	*Anoecia corni* (Fab.)
Ornamental Chokecherry	*Aphis cerasifoliae* Fitch
Creek Dogwood	*Aphis cornifoliae* Fitch
Currant, Gooseberry	*Aphis ribi-gilletei* Knowl.
Snowball Bush, Fig	*Aphis rumicis* L.
Gooseberry	*Aphis varians* Patch
Snowberry	*Brevicoryne symphoricarpi* (Thos.)
Currant	*Cryptomyzus ribis* L.
Plum	*Hyalopterus pruni* (Geoff.)
Currant	*Kakimia cynosbati* (Oest.)
Cherry	*Myzus cerasi* Fab.
Red and White Ash	*Prociphilus corrugatans* (Sirrine)
Oregon Ash	*Prociphilus fraxinifolii* (Riley)
Snowberry, Honeysuckle	*Rhopalosiphum melliferum* (Hottes)

Photo 51: Plum leaves curled by aphids (late June).

GALLS OF THE DRYLAND SHRUBS EAST OF THE CASCADE MOUNTAINS

Compared to western Oregon and Washington, there is an abrupt and dramatic change in the types of plants found east of the Cascade mountains. With different plants come different types of galls. The dryland shrubs that are abundant east of the Cascades—sagebrush (*Artemisia*), rabbit-brush (*Chrysothamnus*), greasewood (*Atriplex*), horse-brush (*Tetradymia*), and juniper (*Juniperus*)—bear their own distinctive types of galls.

There are some interesting points to keep in mind about these galls. One is their abundance. A quick walk through a patch of sagebrush or rabbit-brush at any time of year should convince you that galls are plentiful on the leaves and stems of these plants. There are even reports of insect galls on the below-ground stems of sagebrush. You will also find many different types of galls on dryland plants. These two features, abundance and diversity, are highlighted in photo 52, which shows three different midge (cecidomyiid) galls on a single rabbit-brush twig.

The classification of rabbit-brush plants—the separation of these plants into species or subspecies—is difficult. Recently, however, rabbit-brush galls have been used to distinguish subspecies of the plant. Different plant subspecies bear distinctively different kinds of galls formed by different species of midges.

Photo 52: Three different types of midge galls (arrows) caused by three different species of cecidomyiids occur on a single stem of gray rabbit-brush (*Chrysothamnus nauseosus* (Pall.) Britt.). The gall on the far left is very common, and is rather complex in design; a single larval capsule is surrounded by hairs, which in turn are surrounded by swollen stem tissues.

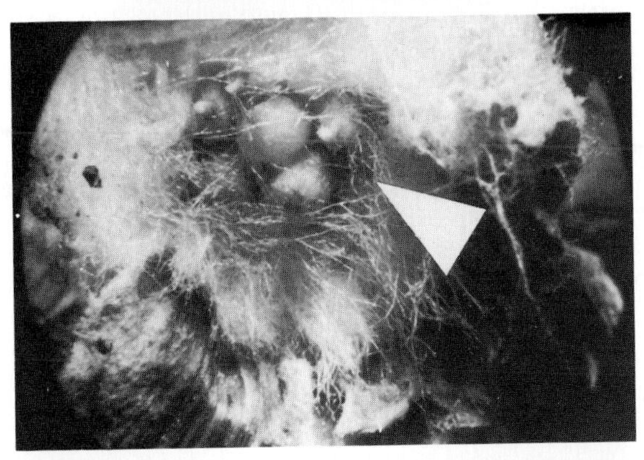

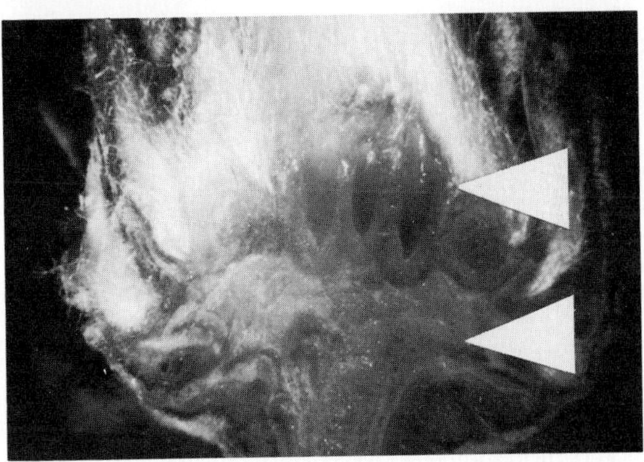

Photo 53-55. A midge (cecidomyiid) gall on gray rabbit-brush. The gall looks like a flower head that has gone to seed. Note the white hairs in the center of the "flower head" (arrow, photo 53). If these hairs are removed, you will see the tops of small egg-shaped larval capsules (arrow, photo 54). A cutaway view (photo 55) shows larval capsules (upper arrow) resting like seeds on a flared base (lower arrow).

Apparently the midges are very discriminating in their choice of host plant, and we can thus use them as plant classifiers.

Many of the galls on dryland shrubs are of interesting design. For example, many are covered or filled with white cottony hairs (photos 53-56). Perhaps the hairs insulate the gall former against quick changes in temperature or against the desiccating sun and winds. Some of the galls resemble either small artichokes or flower heads that have gone to seed (photos 57-59).

The most common leaf gall on sagebrush is spongy and globular (1 to 4 cm in diameter) and is caused by the gall midge (cecidomyiid) *Diarthronomyia artemisiae* Felt. A much less apparent gall in the drylands is found in the seed pods of wild lupine. A callus (i.e., gall) replaces some or all of the seeds within the pod and the gall former, a weevil larva (*Tychius* sp.), feeds on the callus (photo 60).

Interestingly, very few cynipid wasp galls or aphid galls occur on dryland shrubs. Instead, most of the galls are caused by flies—the cecidomyiids (gall midges) and the tephritids (fruit-flies). Moths and mites also cause a few galls on dryland shrubs (photo 61).

Despite the abundance and wide geographical range of many dryland shrub galls, relatively little is known about them, although some were studied as early as 1893. Many of these gall-formers are undescribed, and only a few have been studied in any detail.

Photo 56: A flower-like midge (cecidomyiid) gall on *Artemisia tridentata* Nutt. Internally this gall is very similar in design to the gall on gray rabbit-brush shown in photos 53-55.

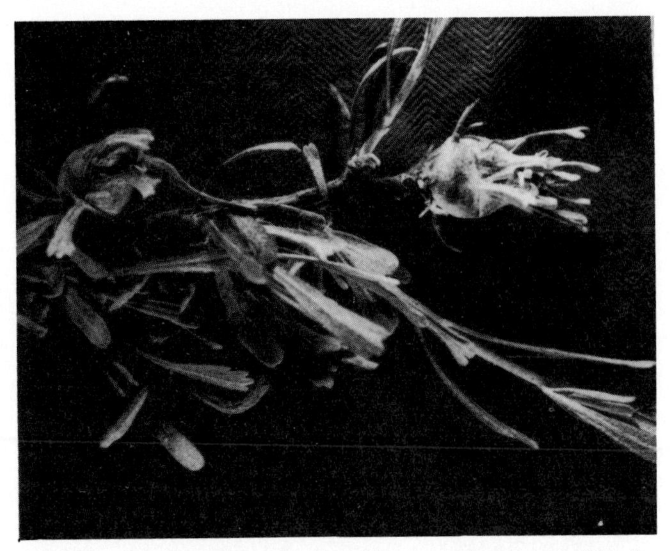

Photo 57: A common artichoke-like gall on *A. tridentata* caused by the tephritid fly, *Eutreta diana* (O.S.).

Photo 58: A red-and-green-spotted artichoke gall on greasewood caused by a gall midge (cecidomyiid).

For more information on dryland shrub galls see:

Fronk, W. D., A. A. Beetle, and D. G. Fullerton. "Dipterous Galls on *Artemisia tridentata* Complex and Insects Associated with Them." *Annals of the Entomological Society of America* 57 (1964): 575-577.

Jones, R. G. "Ecology of *Rhopalomyia* and *Diarthronomyia* Gall Midges (Diptera: Cecidomyiidae) on Sagebrush, *Artemisia* spp., in Idaho." Ph.D. Dissertation. University of Idaho, 1971.

Photo 59: A cluster of artichoke galls on juniper caused by gall midges (cecidomyiids).

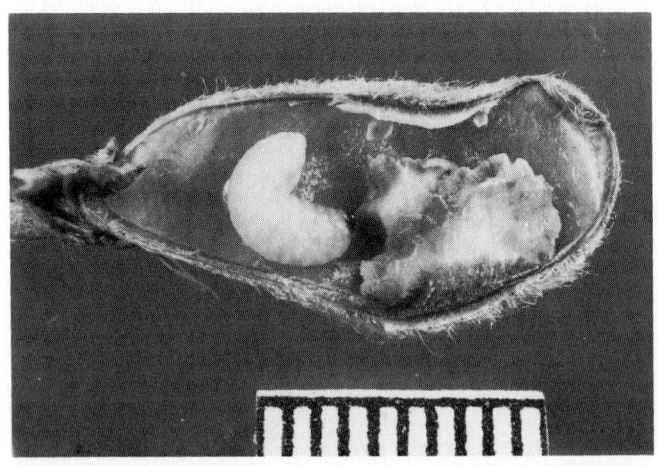

Photo 60: A cutaway view of a galled pod of wild lupine (*Lupinus*) showing the callus that has replaced the seeds, and the gall-forming beetle larva (*Tychius*) as it feeds on the callus. Scale in mm.

McArthur, E. D., C. F. Tiernan, and B. L. Welch. "Subspecies specificity of gall forms on *Chrysothamnus nauseosus.*" *Great Basin Naturalist* 39 (1979): 81-87.

Wangberg, J. K. "Biology of the Tephritid Gall-formers (Diptera: Tephritidae) on Rabbit-brush, *Chrysothamnus* spp., in Idaho," Ph.D. Dissertation. University of Idaho, 1976.

Photo 61: This gall on spiny horse-brush (*Tetradymia spinosa* H. and A.) is caused by the gelechiid moth, *Gnorimoschema tetradymiella* Busck. The gall is of simple design with thickened walls and a long central larval cavity. A similar, slightly more rounded gall occurs on *T. glabrata* Gray.

GLOSSARY

Agamic generation—A generation of female gall wasps. The females require no mating to lay viable eggs.

Alternation of generation—In gall wasps, the cycling between an agamic and a bisexual generation.

Arthropods—The large group of backboneless animals that have jointed appendages such as legs, and also have a hard, chitinous shell. Insects and mites are arthropods.

Bisexual generation—A generation of both male and female gall wasps. Females must mate before they lay viable eggs.

Crucifers—Plants in the cabbage family.

Deutogyne—A specialized, overwintering female eriophyoid mite.

Erineum (erinea, plural)—A dense patch of hairs on leaves. A felt-like or velvety pad usually caused by mites or fungi.

Gall—An abnormal proliferation, growth, or swelling on a plant caused by another organism.

Inquilines—Insects and mites that live as guests in a gall. They do not form the gall, but often live alongside the gall-former.

Larva (larvae, plural)—The immature worm-shaped stage of insects that undergo complete metamorphosis. Grubs, caterpillars, maggots.

Nymph—The immature stage of insects that undergo incomplete metamorphosis. A nymph looks like a small wingless version of the adult.

Pupa (pupae, plural)—An inactive life stage that occurs between the larval and adult stage in flies, wasps, moths, and beetles.

Pupate—To become a pupa.

Stem-mother aphid—The female aphid which, by herself, often forms a gall and then populates it with her offspring.

Sternal spatula—A small brown Y- or T-shaped structure on the undersurface of and near the head of older midge (cecidomyiid) larvae.

Witches broom—A type of gall. At a spot on a plant, many short branchlets arise. Typically the abnormality resembles the head of a broom among the branches of a tree. Mites and fungi are often responsible for broom formation.